A2-Level
Biology
AQA B
The Revision Guide

Editor:
Rachel Selway.

Contributors:
Gloria Barnett, Ellen Bowness, Wendy Butler, Martin Chester, James Foster,
Julian Hardwick, Derek Harvey, Kate Houghton, Simon Little, Kate Redmond, Katherine Reed,
Adrian Schmit, Emma Singleton, Jennifer Underwood.

Proofreader:
James Foster.

Published by Coordination Group Publications Ltd.

ISBN: 1 84146 395 7
Groovy website: www.cgpbooks.co.uk
Jolly bits of clipart from CorelDRAW
Printed by Elanders Hindson, Newcastle upon Tyne.

Contents

Core Content

Section One — Energy Supply

Section Two — Control, Coordination & Homeostasis

Section Three — Inheritance

Section Four — Environment

Optional Content

Section Five — Applied Ecology

Section Six — Microbes and Disease

Section Seven — Behaviour & Populations

ATP and Energy Supply

All animals and plants need energy for life processes and also for reading books like this. This stuff is pretty tricky and we're diving in at the deep end, so hang on...

Biological processes need **Energy**

Cells need **chemical energy** for biological processes to occur. Without this energy, these processes would stop and the animal or plant would just **die**... not good.

Energy is needed for **biological processes** like:
- active transport
- muscle contraction
- maintenance of body temperature
- reproduction and growth

Plants need energy for **metabolic reactions**, like:
- photosynthesis
- taking in minerals through their roots

ATP carries **Energy** around

It only **gets worse** from here on in for the rest of the section. But **don't worry**, ATP didn't make any sense to me at first — it just clicked after many **painful hours** of reading dull books and listening to my teacher going on and on. Here goes...

1) ATP (**adenosine triphosphate**) is a **small water-soluble** molecule that is easily transported around cells.
2) It's made from the nucleotide base **adenine**, combined with a **ribose sugar** and **three phosphate groups**.
3) ATP is a **phosphorylated nucleotide** — this means it's a nucleotide with extra phosphate groups added.
4) ATP **carries energy** from **energy-releasing** reactions to **energy-consuming** reactions.

How ATP carries energy:

1) ATP is **synthesised** from **adenosine diphosphate (ADP)** and an **inorganic phosphate** group using the energy produced by the **breakdown of glucose**. The enzyme **ATPsynthase** catalyses this reaction.
2) ATP **moves** to the part of the cell that requires energy.
3) It is then **broken down** to ADP and **inorganic phosphate** and **releases chemical energy** for the process to use. **ATPase** catalyses this reaction.
4) The ADP and phosphate are **recycled** and the process starts again.

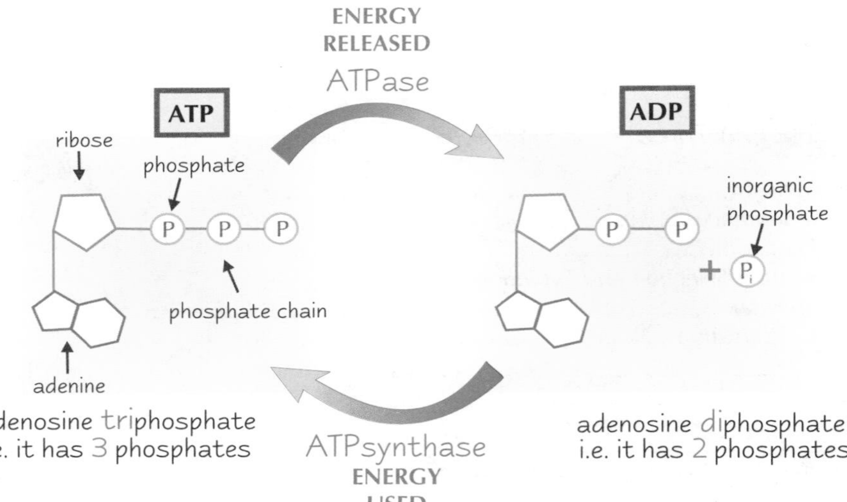

Cells Release Energy (to make **ATP**) by **Respiration**

Cellular respiration is the process where cells **break down glucose**, it produces carbon dioxide and water and releases **energy**. The energy is used to **produce ATP** from ADP and P_i. There are two types of respiration:

1) **Aerobic respiration** — respiration **using oxygen**.
2) **Anaerobic respiration** — respiration **without oxygen**. Both types produce ATP.

You need to learn the summary equation for **aerobic respiration**.

$$C_6H_{12}O_6\text{ (glucose)} + 6O_2 \longrightarrow 6CO_2 + 6H_2O + \text{Energy}$$

ATP and Energy Supply

Aerobic Respiration takes place in the Mitochondria of the Cell

1) **Mitochondria** are present in all **eukaryotic** (i.e. plant, animal, fungi and protoctist) cells. They're 1.5 to 10 µm long.
2) Cells that use lots of energy, e.g. **muscle cells**, **liver cells** and the middle section of **sperm**, have lots of mitochondria.
3) The **inner membrane** of each mitochondrion is folded into **cristae** — structures that increase surface area.
4) **ATP** is produced via the **stalked particles** on the cristae of the inner mitochondrial membrane, in a stage called the **electron transport chain** (see page 7).
5) The **Krebs Cycle** (page 6) takes place in the **matrix** of mitochondria.

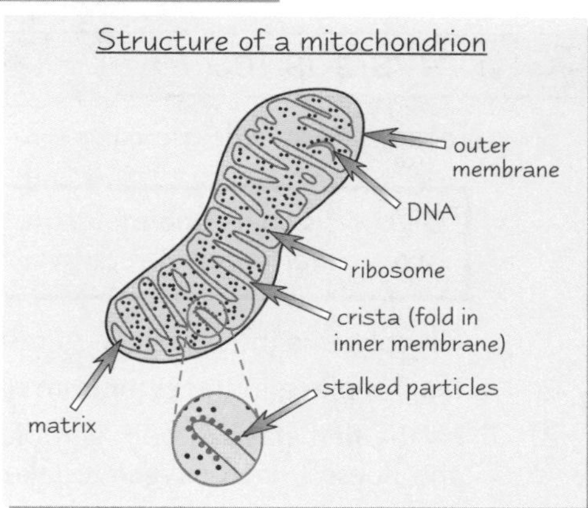

Respiration is a Metabolic Pathway

There are some pretty confusing technical terms about reactions in this section.
If you do chemistry you'll be laughing — if not, you'd better concentrate:

- **Metabolic pathway** — a **series** of **small reactions**, e.g. respiration or photosynthesis, controlled by enzymes.
- **Catabolic reactions** — **breaking large molecules** into **smaller ones** using enzymes, e.g. breaking down glucose in respiration.
- **Anabolic reactions** — **combining smaller molecules** to make **bigger ones** using enzymes.
- **Phosphorylation** — **adding phosphate** to a molecule, e.g. ADP is phosphorylated to ATP.
- **Hydrolysis** — the **splitting** of a molecule using **water**.
- **Photolysis** — the **splitting** of a molecule using **light** energy.

Redox reactions — reactions that involve **oxidation** and **reduction**.

1) If something is **reduced** it has **gained electrons**, and may also have lost oxygen or gained **hydrogen**. If something is **oxidised** it has **lost electrons**, and may also have gained oxygen or lost **hydrogen**.
2) Oxidation of one thing always involves reduction of something else.
3) The enzymes that catalyse redox reactions are called **oxidoreductases**.
4) Respiration and photosynthesis are riddled with redox reactions.

Oxidation	Reduction
electrons are lost	electrons are gained
oxygen is added	oxygen is lost
hydrogen is lost	hydrogen is gained

One way to remember electron movement is "OILRIG" = Oxidation Is Loss of e^-, Reduction Is Gain of e^-.

Practice Questions

Q1 How is energy released from ATP?

Q2 Write down five metabolic processes in animals which require energy.

Q3 What is the purpose of the cristae in mitochondria?

Q4 What are oxidoreductases?

Exam Questions

Q1 What is the connection between phosphate and the energy needs of a cell? [2 marks]

Q2 ATP is a small, water-soluble molecule which can be rapidly and easily converted back into ADP if ATPsynthase is present. Explain how these features make ATP suitable for its function. [3 marks]

I've run out of energy after that little lot...

You really need to understand what ATP is, because once you start getting bogged down in the complicated details of respiration and photosynthesis, at least you'll understand why they're important and what they're producing. It does get more complicated on the next few pages, so take your time to understand the basics before you turn the page.

Glycolysis and the Link Reaction

You can split the process of respiration into four parts — that way you don't have to swallow too many facts at once. The first bit, glycolysis, is pretty straightforward.

Glycolysis** is the **First Stage** of **Respiration

So, to recap... most cells use carbohydrates, usually glucose, for respiration.

Glycolysis splits **one molecule** of glucose into **two** smaller molecules of **pyruvate**.

Glucose is a hexose (6-carbon) molecule.
Pyruvate is a triose (3-carbon) molecule.
Pyruvate is also known as pyruvic acid.

Respiration Map

Glycolysis — *You are here*
↓
Link Reaction
↓
Krebs Cycle
↓
Electron Transport Chain

1) Glycolysis is the first stage of respiration (see the map to the right).
2) It takes place in the **cytoplasm** of cells.
3) It's the **first stage** of both aerobic and anaerobic respiration, and **doesn't need oxygen** to take place — so it's **anaerobic**.

*There are **Two Stages** of **Glycolysis** — **Phosphorylation** and **Oxidation***

1 Stage One — Phosphorylation

1) Glucose is **phosphorylated** by adding 2 **phosphates** from 2 molecules of ATP.
2) Glucose is split using water (**hydrolysis**).
3) 2 molecules of **triose phosphate** are created and 2 molecules of ATP are used up.

2 Stage Two — Oxidation

1) The triose phosphate is **oxidised** (loses hydrogen), forming **two** molecules of **pyruvate**.
2) **Coenzyme NAD$^+$** collects the hydrogen ions, forming **2 reduced NAD** (**NADH + H$^+$**).
3) **4 ATP** are produced, but 2 were used up at the beginning, so there's a **net gain** of **2 ATP**.

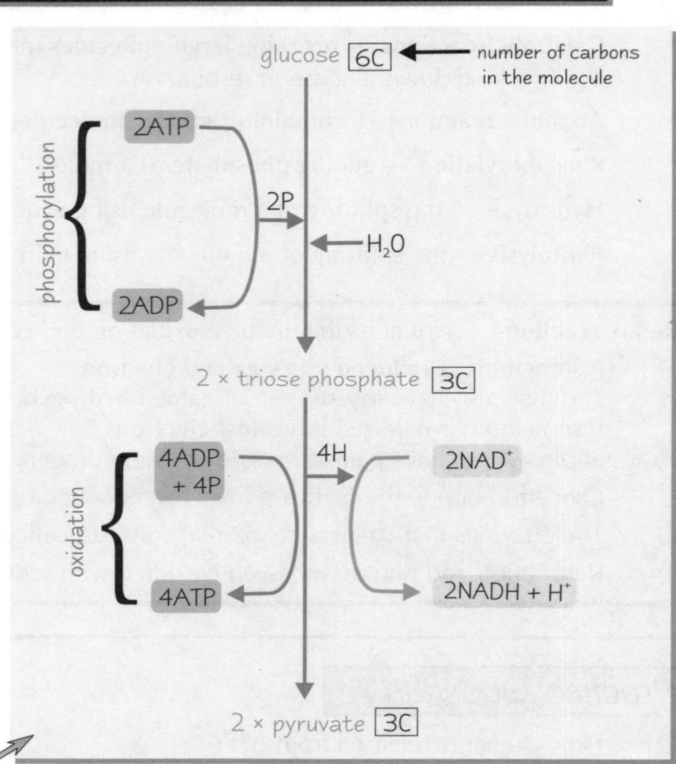

A coenzyme is a helper molecule that carries chemical groups or ions about, e.g. NAD removes H$^+$ and carries it to other molecules.

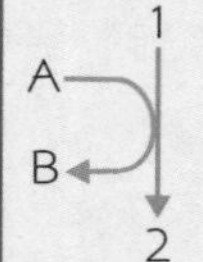

These arrows in diagrams just mean that A goes into the main reaction and is converted to B. A will normally release or collect something from molecule 1, e.g. hydrogen or phosphate.

A triose phosphate is just a simple 3-carbon sugar with a phosphate group attached. Different books use different names, but this is the easiest to remember.

Next** in **Aerobic Respiration...

1) The **2** molecules of **reduced NAD** go to the **electron transport chain** (see page 7).
2) The **two pyruvate** molecules go into the matrix of the **mitochondria** for the **link reaction** (a small reaction that **links** glycolysis to the next stage, the **Krebs cycle**). It's so exciting I bet you can't wait...

Glycolysis and the Link Reaction

The Link Reaction converts Pyruvate to Acetylcoenzyme A

The link reaction is fairly simple and goes like this:

1) One **carbon atom** is removed from pyruvate in the form of CO_2.
2) The remaining **2-carbon molecule** combines with **coenzyme A** to produce **acetylcoenzyme A (acetyl CoA)**.
3) Another oxidation reaction occurs when NAD^+ collects more **hydrogen ions**. This forms **reduced NAD** ($NADH + H^+$).
4) **No ATP** is produced in this reaction.

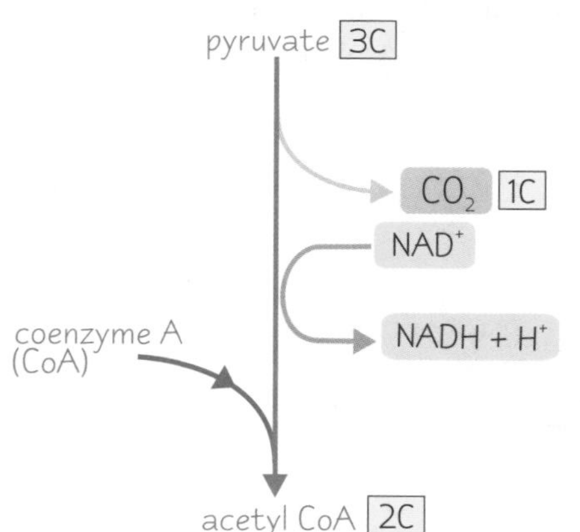

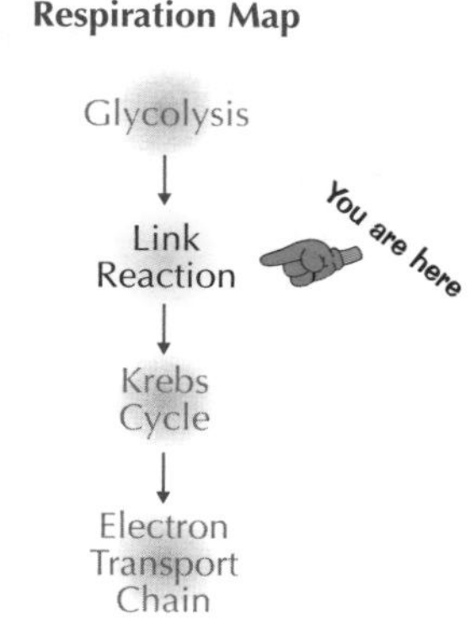

The Link Reaction occurs Twice for every Glucose Molecule

1) For each **glucose molecule** used in glycolysis, **two pyruvate** molecules are made.
2) But the **link reaction** uses only **one pyruvate** molecule, so the **link reaction** and the **Krebs cycle** happen **twice** for every glucose molecule which goes through glycolysis.

The Products of the Link Reaction go to the Krebs Cycle and the ETC

So for each glucose molecule:

- Two molecules of **acetylcoenzyme A** go into the Krebs cycle (see next page).
- Two **carbon dioxide molecules** are released as a waste product of respiration.
- Two molecules of **reduced NAD** are formed and go into the **electron transport chain** (which is covered on the next two pages).

Practice Questions

Q1 What do the terms hydrolysis and phosphorylation mean?

Q2 Why is there only a net gain of 2 ATP during glycolysis?

Q3 How many carbon atoms are there in triose phosphate and in pyruvate?

Q4 Where is acetyl CoA formed?

Exam Questions

Q1 Describe simply how a 6-carbon molecule of glucose can be changed to pyruvate. [5 marks]

Q2 Describe what happens in the link reaction. [4 marks]

Acetyl Co-what?

It's all a bit confusing, but you need to know it, so it's worth taking a bit of time to break it down into really simple chunks. Don't worry too much if you can't remember all the little details straight away. If you can remember how it starts and what the products are, you're getting there. You'll get the hang of it all eventually, even if it seems hard right now.

Krebs Cycle and the Electron Transport Chain

And now we have the third and fourth stages of the respiration pathway. Keep it up — you're nearly there.

*The **Krebs Cycle** is the **Third Stage** of Aerobic Respiration*

The Krebs cycle involves a series of **oxidation reactions** which take place in the **matrix** of the mitochondria. The cycle happens once for each pyruvate molecule made in glycolysis, so it goes round twice for every glucose molecule that enters the respiration pathway.

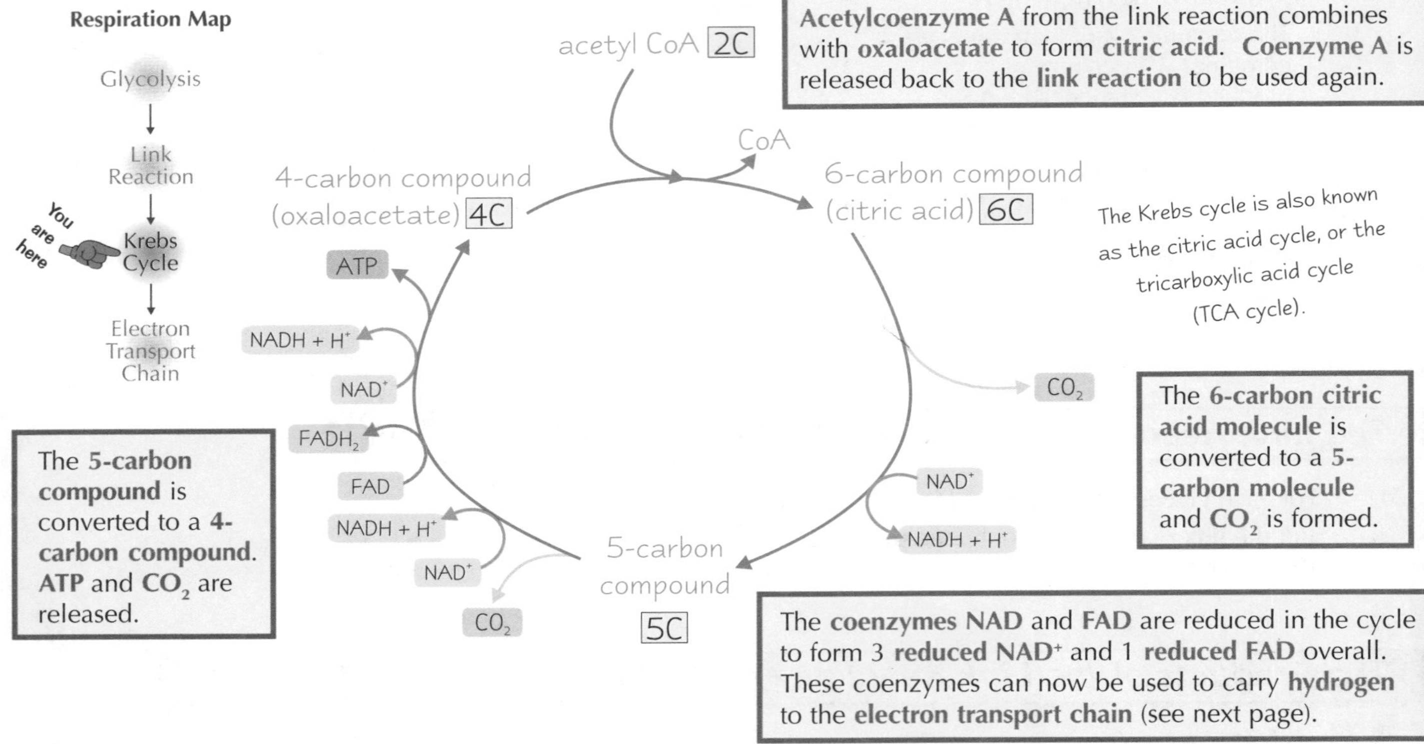

Products of the Krebs Cycle are used in the Electron Transport Chain

Some products are **reused**, some are **released** and others are used for the **next stage** of respiration:

- One **coA** is **reused** in the next **link reaction**.
- **Oxaloacetate** is **regenerated** so it can be **reused** in the next **Krebs cycle**.

- Two **carbon dioxide** molecules are released as a **waste product** of respiration.
- One molecule of **ATP** is made per turn of the cycle — directly from these chemical reactions.

- **Three reduced NAD** and **one reduced FAD** co-enzymes are made and carried forward to the **electron transport chain**.

*The **Electron Transport Chain** is the **Final Stage** of Aerobic Respiration*

Before we get too bogged down in all the details, here's what the electron transport chain is all about:

Reduced NAD and **reduced FAD** from the previous stages are used in this final stage. Its purpose is to **transfer** the **energy** from molecules made in glycolysis, the link reaction and the Krebs cycle to ADP. This forms **ATP**, which can then deliver the energy to parts of the cell that need it.

The electron transport chain is where **most of the ATP** from respiration is produced. In the whole process of aerobic respiration, **32 ATP molecules** are produced from one molecule of glucose: 2 ATP in glycolysis, 2 ATP in the Krebs cycle and 28 ATP in the electron transport chain.

The electron transport chain also **reoxidises NAD and FAD** so they can be reused in the previous steps.

Krebs Cycle and the Electron Transport Chain

The Electron Transport Chain produces lots of ATP

The **electron transport chain** uses the molecules of **reduced NAD** and **reduced FAD** from the previous three stages to produce **28 molecules of ATP** for every molecule of glucose.

1) **Hydrogen atoms** are released from **NADH + H^+** and **$FADH_2$** (as they are oxidised to NAD^+ and FAD). The H atoms **split** to produce **protons (H^+)**, and **electrons (e^-)** for the chain.

2) The **electrons** move along the electron chain (made up of three **electron carriers**), losing energy at each level. This energy is used to **pump** the **protons** (H^+) into the space **between** the inner and outer **mitochondrial membranes** (the **intermembrane space**).

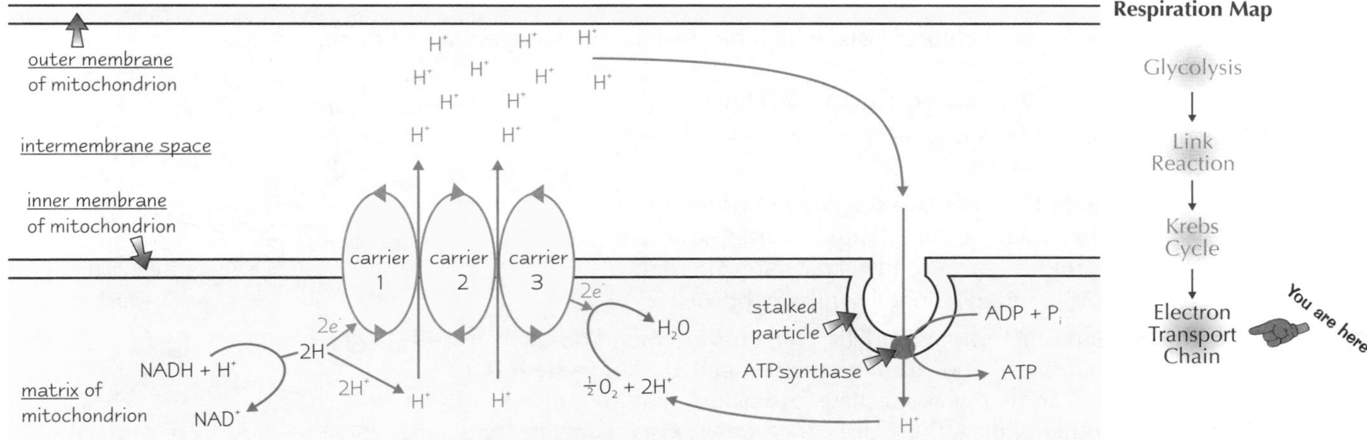

3) The **concentration** of protons is higher in the intermembrane space than in the mitochondrial matrix, so an **electrochemical gradient** exists.

4) The **protons** then move back through the inner membrane **down** the **electrochemical gradient**, through specific channels on the **stalked particles** of the **cristae** — this drives the enzyme **ATPsynthase.** By 'spinning like a motor', this enzyme supplies **electrical potential energy** to make **ATP** from ADP and inorganic phosphate.

5) The **protons** and **electrons** recombine to form **hydrogen**, and this combines with **molecular oxygen** (from the blood) at the end of the transport chain to form **water**. Oxygen is said to be the final **electron acceptor**.

The **synthesis of ATP** as a result of the energy released by the electron transport chain is called **oxidative phosphorylation**.

This is how the **electron transport chain** produces **28** molecules of **ATP** from **1** molecule of **glucose**:

- 1 turn of the Krebs cycle produces **4** molecules of **reduced NAD** (including **1** from the **link reaction**) and **1** of **reduced FAD**.
- 2 molecules of **pyruvate** enter the Krebs cycle for **each** molecule of **glucose**, so overall **8 NAD^+** and **2 FAD** are reduced.
- **2 reduced NAD** are also produced from the first part of respiration, **glycolysis** (see p. 4).
- Each **reduced NAD** can produce **2.5 ATP**, and each **reduced FAD** can produce **1.5 ATP**.
- So: 8 reduced NAD + 2 reduced NAD = **10** reduced NAD. **10 × 2.5 = 25 ATP**. **2 reduced FAD × 1.5 = 3 ATP.** In total, 25 + 3 = **28 molecules of ATP.**

(There are also **2 ATP** produced by **glycolysis**, and **2** for each molecule of glucose in the **Krebs cycle** = **32 ATP** produced in total by **respiration**.)

Practice Questions

Q1 How many molecules of CO_2 are made in one turn of the Krebs cycle?

Q2 Name the two coenzymes that are reduced in the Krebs cycle.

Q3 Which molecule finally accepts the electrons passed down through the electron transport chain?

Exam Question

Q1 Calculate the number of ATP molecules that are produced by aerobic respiration from one molecule of glucose. Show your working in detail. [14 marks]

That's all folks... until photosynthesis...

Phew, this biochemistry stuff is tough going. The key to learning this stuff is to learn the big facts first — glycolysis, link reaction, Krebs cycle, electron transport chain. Once you know what the main parts are, which order they come in and roughly what happens at each stage, you stand some chance of learning the more detailed stuff.

Photosynthesis

Don't worry if this seems hard at first. Read it through carefully a couple of times, and it'll start to make sense.

Photosynthesis *happens in the* ***Chloroplasts***

Here's the overall equation for photosynthesis. Hopefully it'll look pretty familiar. When you were doing your GCSEs, this little equation was all you had to worry about, but those days are long gone, my friend.

$$6CO_2 + 6H_2O + \text{Energy} \xrightarrow{\text{chlorophyll}} C_6H_{12}O_6 + 6O_2$$

Photosynthesis happens in the **chloroplasts**, which means you need to learn about them:

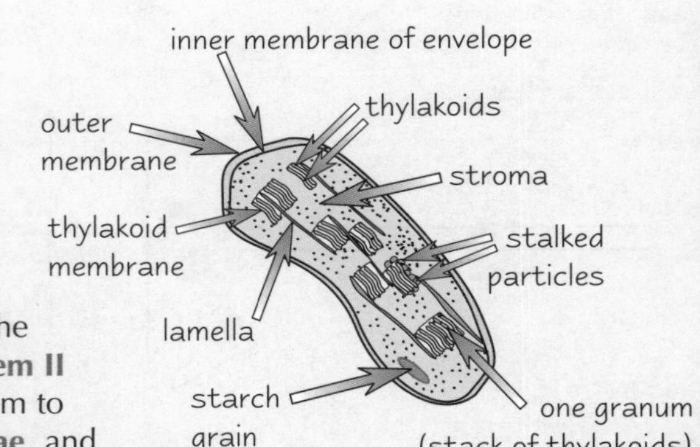

1) Chloroplasts are usually about **5 μm** in diameter.
2) They're surrounded by a double membrane called the **chloroplast envelope**.
3) **Thylakoids** (fluid-filled sacs) are stacked up inside the chloroplast into structures called **grana**. These structures have a large surface area. The thylakoids are where the **light-dependent reaction** of photosynthesis occurs.
4) **Chlorophyll** and other photosynthetic pigments are found on the **thylakoid membranes**. They form a complex called **photosystem II** (see next page). Some thylakoids have **extensions** that join them to thylakoids in other grana. These are called **inter-granal lamellae**, and they're the sites of **photosystem I** (see next page).
5) The thylakoids are embedded in a gel-like substance called the **stroma**. The stroma is where the **light-independent reaction** of photosynthesis (called the **Calvin cycle**) happens. It contains **enzymes** (for the Calvin cycle), **sugars** and **organic acids**.
6) Carbohydrates produced by photosynthesis and not used straight away are stored as **starch grains** in the **stroma**.

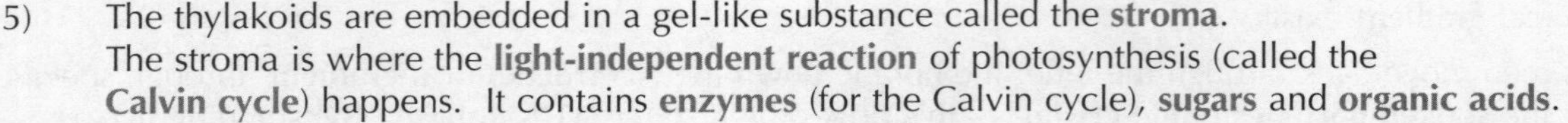

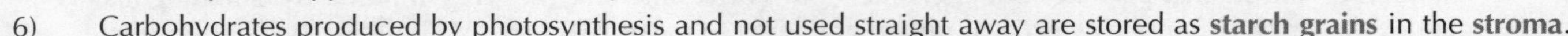

Photosynthesis *can be Split into* ***Two Stages***

Nowadays you need to know that photosynthesis consists of two **stages**:

1) The **light-dependent reaction** (which, as the name suggests, needs **light energy**) takes place in the **thylakoid membranes** of the chloroplasts. Light energy is absorbed by pigments in the **photosystems** (see below), and used to provide the energy for the next stage — the light-independent reaction. There are **two different reactions** going on in this next stage — **cyclic photophosphorylation** and **non-cyclic photophosphorylation**. The plant can **switch between** the two, depending on whether it needs **reduced NADP** or just **ATP** (see below).

2) The **light-independent reaction** or **Calvin cycle** (which, as the name suggests, doesn't use light energy) happens in the **stroma** of the chloroplast. The **ATP** and the **reduced NADP** molecules that were made in the light-dependent reaction supply the **energy** to make **glucose**. See pages 10-11 for more on the Calvin cycle.

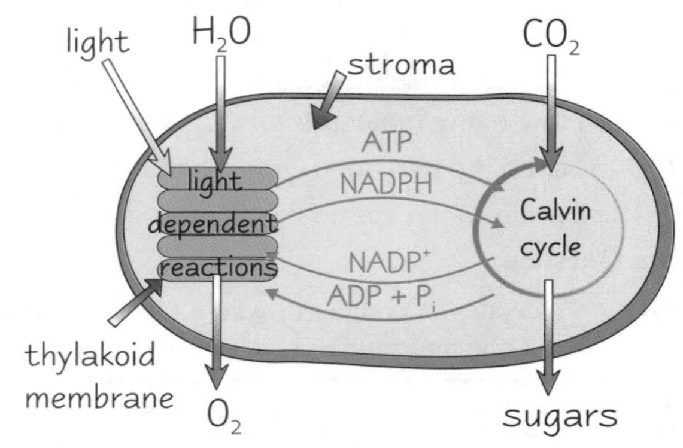

The diagram shows how the two different reactions, light-dependent and light-independent, fit together in the chloroplast.

Photosynthesis

Photosystems I and II capture Light Energy

1) **Photosystems** are made up of **chlorophyll a**, **accessory pigments** (like **chlorophyll b** and **carotenoids**) and **proteins**. The proteins hold the pigment molecules in the best positions for **absorbing** light energy and **transferring** this energy to the **reaction centre** of the photosystem.
2) The **reaction centre** is a particular **chlorophyll molecule** called a **primary pigment**. The energy from absorbing light is passed from one **accessory pigment** to another until it reaches this **primary pigment**.
3) The energy is then used to **excite** pairs of **electrons** in the reaction centre pigment. The electrons move up to a higher **energy level**, ready to be used in the **light dependent reactions**.
4) There are **two** different photosystems used by plants to capture light energy. **Photosystem I** (or PSI) uses a chlorophyll molecule that absorbs light at wavelength **700 nm** in its reaction centre. **Photosystem II** (PSII) uses a chlorophyll molecule that absorbs light best at around **680 nm** in its reaction centre.

The Light-Dependent Reaction makes ATP in Photophosphorylation

The **energy** captured by the photosystems is used for **two** main things:

1) Making **ATP** from **ADP** and **inorganic phosphate** (**phosphorylation**). It's called **photo**phosphorylation here, as it uses **light**.
2) Splitting **water** into H^+ ions and **oxygen**. This is called **photolysis**, because the splitting (lysis) is caused by light energy (photo). Photolysis is covered on the next page.

Photophosphorylation involves the **excited electrons** in the reaction centre of a photosystem being passed to a special molecule called an **electron acceptor**. These electrons are then passed along a **chain** of other electron carriers, each at a slightly **lower energy level** than the one before, so that the electrons **lose energy** at every stage in the chain. The energy given out is used to add a phosphate molecule to a molecule of ADP — and this is photophosphorylation.

You don't need to know the actual mechanism in detail, but it's very similar to how ATP is produced in **respiration**, involving the flow of **hydrogen ions** (H^+) through **stalked particles** (which you can read all about on page 7). The H^+ ions used in photosynthesis come from the **photolysis of water**, which is covered on the next page.

The difference between **cyclic** and **non-cyclic photophosphorylation** is in what happens to those electrons that have been moving through the chain of carriers. Non-cyclic photophosphorylation is explained on the next page.

Cyclic Photophosphorylation just produces ATP

Cyclic photophosphorylation only uses **photosystem I**. It's called cyclic photophosphorylation because the electrons from the chlorophyll molecule are simply **passed back** to it after they've been through the chain of carriers — i.e. they're **recycled** and can be used repeatedly by the same molecule. This **doesn't** produce any **reduced NADP** (NADPH + H^+), but there **is** enough energy to make **ATP**. This can then be used in the **light-independent reaction**.

Practice Questions

Q1 What is the full equation for photosynthesis?

Q2 Where in the chloroplast does the light-independent reaction of photosynthesis happen?

Q3 What is the reaction centre of a photosystem?

Q4 What two main things is the light energy captured by photosystems used for?

Exam Question

Q1 a) Where precisely in the plant does the light-dependent stage occur? [1 mark]

b) Which two compounds produced in the light-dependent stage are used in the light-independent stage? [2 marks]

Photophosphorylate that, if you can...

*If you're feeling filled with despair as you read this tip, well, don't. You **will** understand this, don't give up. I guarantee it'll seem clearer every time you go through it, until at last you're left wondering what all the fuss was about. By the way, don't be put off when it says protons instead of hydrogen ions. That always confused me, but they mean the same thing.*

Photosynthesis

Here's the rest of the stuff on photosynthesis... It ain't pretty, but it's here and it's all yours. Have fun and learn it well.

Non-cyclic Photophosphorylation produces ATP, NADPH and Oxygen

Non-cyclic photophosphorylation uses both **PSI** and **PSII**. It involves **photolysis**, which is the splitting of **water** using light energy. Photolysis only happens in **PSII**, because only PSII has the right **enzymes**.

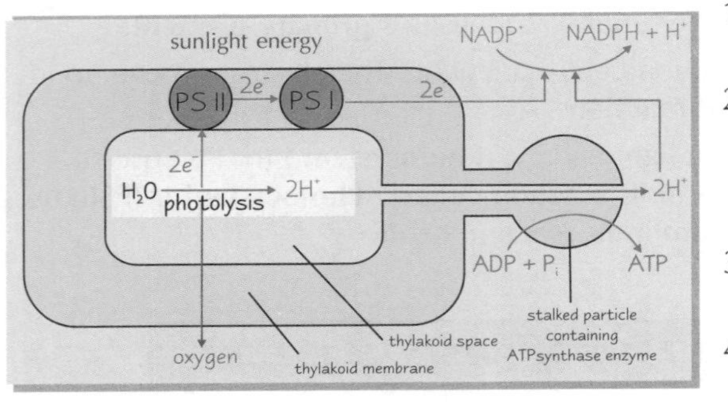

1) Light energy enters **PSII** and is used to move electrons to a **higher energy level**.
2) The electrons are passed along the chain of electron carriers to **photosystem I**. Most of the energy lost by the electrons during this process is used to make **ATP** (like in cyclic photophosphorylation).
3) Light energy is absorbed by PSI, which excites the electrons **again** to an **even higher energy level**.
4) The electrons are passed to a **different electron acceptor**, and **don't** return to the chlorophyll.
5) For the chlorophyll to keep working, the electrons have to be replaced from **somewhere else** — so they're taken from a molecule of **water** (water is the electron donor). This makes the water molecule split up into **protons** (H^+) and **oxygen**.
6) The **protons** (H^+) from the water molecule combine with the **electrons** currently with the second electron acceptor to give **hydrogen atoms**. These are used to react with a substance called **NADP** to produce **NADPH** and **H^+**. These are needed for the **light-independent reaction**.

So when electrons move back from the **second electron acceptor** to the chlorophyll molecule, that's **cyclic** photophosphorylation. If they don't, and the replacement electrons come from **water** instead, that's **non-cyclic**.

	cyclic photophosphorylation	non-cyclic photophosphorylation
photosystem	I	I and II
what's needed	light, ADP, inorganic phosphate	light, water, NADP, ADP, inorganic phosphate
what's produced	ATP	ATP, NADPH + H^+, O_2

The Light-Independent Reaction is also called the Calvin Cycle

The Calvin cycle makes **hexose sugars** (sugars with **6 carbons**, like **glucose** and **fructose**) from **carbon dioxide** and a **5-carbon** compound called **ribulose bisphosphate**. It happens in the **stroma** of the chloroplasts. There are a few steps in the reaction, and it needs **energy** and **H^+ ions** to keep the cycle going. These are provided by the products of the **light-dependent reaction**, **ATP** and **reduced NADP** (**NADPH + H^+**).

The diagram shows what happens at each stage in the cycle. The numbers in brackets (5C, 3C etc.) show how many **carbon atoms** there are in each molecule — the cleverest bit of the cycle is how it turns a **5-carbon** compound into a **6-carbon** one.

1) **CO_2** enters the leaf through the **stomata** and diffuses into the **stroma** of the chloroplast.
2) There it's taken up by **ribulose bisphosphate** (**RuBP**), a **5-carbon** compound. This gives an **unstable 6-carbon** compound, which quickly breaks down into **two** molecules of a **3-carbon** compound called **glycerate 3-phosphate** (**GP**).
3) This reaction is catalysed by the enzyme **ribulose bisphosphate carboxylase** (**rubisco**).
4) **ATP** from the **light-dependent stage** of photosynthesis is now used to provide the energy to turn the **3-carbon** compound, **GP**, into a **different** 3-carbon compound called **triose phosphate**.
5) This reaction also needs **H^+ ions**, which are provided by the **reduced NADP** (**NADPH + H^+**) made in the **light-dependent reaction**.
6) **Two** triose phosphate molecules then **join together** to give **one hexose sugar** (e.g. glucose).

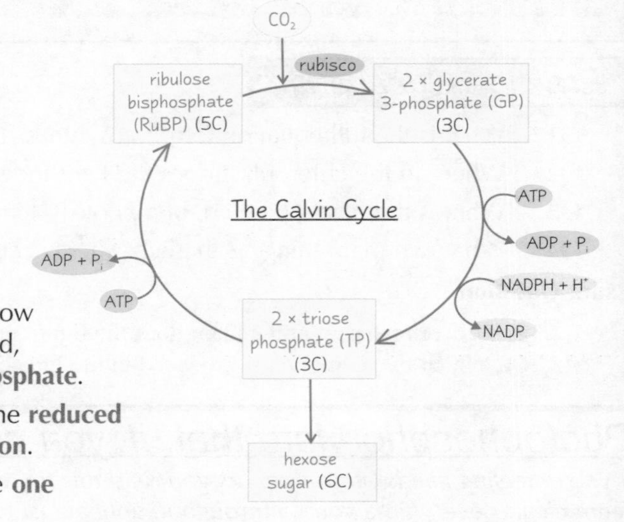

Photosynthesis

The Calvin Cycle needs to turn 6 times to make 1 Glucose Molecule

Five out of every **six** molecules of **triose phosphate** produced in the Calvin cycle are **not** used to make hexose sugars, but to **regenerate** RuBP. Making RuBP from triose phosphate molecules uses the rest of the ATP produced by the light-dependent reaction. This means that the cycle has to happen **six times** just to make one new sugar. The box below shows why.

1) **6 RuBP** and **6 CO_2** molecules (i.e. 6 turns of the cycle) convert a total of **12 glycerate 3-phosphate** molecules into **12 triose phosphate molecules.**
2) **Two** molecules of **triose phosphate** are removed from the cycle and make **one glucose molecule.**
3) **10 triose phosphate** molecules regenerate the RuBP.

This might seem a bit inefficient, but it keeps the cycle going and makes sure that there's always **enough RuBP** there ready to combine with CO_2 taken in from the atmosphere.

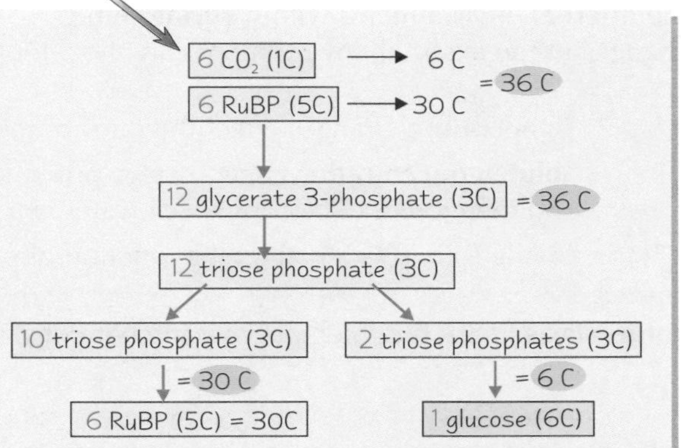

The Calvin cycle is the starting point for making **all** the substances a plant needs — plants can't take in **proteins** and **lipids** like animals can.

- Other **carbohydrates**, like **starch**, **sucrose** and **cellulose**, can be easily made by joining the simple hexose sugars together in different ways.
- **Lipids** are made using **glycerol** synthesised from **triose phosphate**, and **fatty acids** from **glycerate 3-phosphate**.
- **Proteins** are made up of **amino acids**, which are also synthesised from **glycerate 3-phosphate** and nitrate ions.

Practice Questions

Q1 Name three substances that non-cyclic photophosphorylation produces.

Q2 How many carbon atoms are there in a molecule of triose phosphate?

Q3 Name the enzyme that catalyses the reaction between carbon dioxide and ribulose bisphosphate.

Q4 How many CO_2 molecules need to enter the Calvin cycle to make one glucose?

Q5 How is the Calvin cycle involved in making lipids?

Exam Questions

Q1 Which molecule in photosynthesis:

a) is the carbon dioxide acceptor? [1 mark]

b) provides the hydrogen ions to reduce glycerate 3-phosphate? [1 mark]

c) is regenerated in the Calvin cycle? [1 mark]

d) is known as rubisco? [1 mark]

e) is made of 5 carbon atoms? [1 mark]

Q2 Look at the diagram on the right and describe what is happening:

a) between points a and b, [1 mark]

b) between points b and c, [1 mark]

c) at point c. [1 mark]

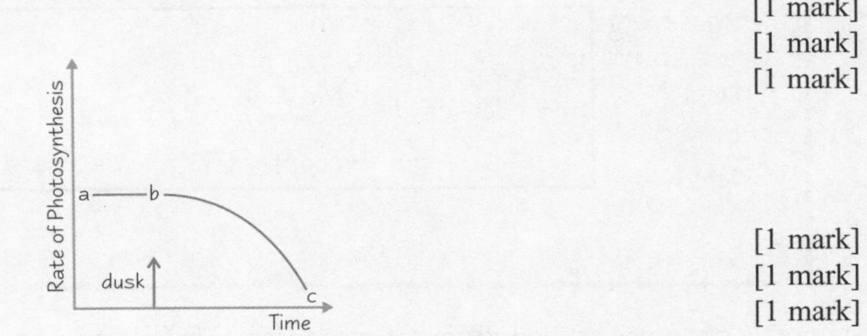

Don't worry — I promise the next section will be a lot easier...

Now don't start sinking into despair again. I know it's a lot to take coming after that last page, but actually this one is probably a bit easier. Learn that cycle on the last page and you're three quarters of the way there. Don't worry too much about learning the maths bit in the box on this page either — as long as you've got the general gist, that's enough.

Homeostasis and Temperature Control

This section is all about keeping things constant inside the body.
Everything has to be carefully balanced — otherwise your body would be totally out of control.

Homeostasis keeps the Internal Environment **Constant**

Homeostasis keeps the **blood** and the **tissue fluid** that surrounds the cells (the **internal environment**) within **certain limits**, so the cells can function normally. **Changes** in the external environment can affect the internal one, which can **damage cells**:

1) **Temperature** changes affect the rates of metabolic reactions, and high temperatures can denature proteins.
2) **Solute concentrations** affect **water potentials** of solutions and therefore the loss or gain of water by cells due to osmosis.
3) Changes in **pH** can affect the function of proteins by changing their shapes.

Homeostasis keeps the internal environment **constant**, avoiding **cell damage**.

A **Homeostatic System** detects a **Change** and **Responds** to it

1) A **receptor detects** a change (the **stimulus**).
2) The receptor communicates with the part of the body that brings about a response (the effector), via the **nervous system** or **hormones**.
3) An **effector** brings about the response. Glands and muscles are effectors.

Negative feedback keeps the internal environment **constant**:

Changes in the environment trigger a response that **counteracts** the changes — e.g. a **rise** in temperature causes a response that **lowers body temperature**.

This means that the **internal environment** tends to stay around a **norm**, the level at which the cells work best.

This only works within **certain limits** — if the environment changes too much, then the effector may not be able to **counteract** it.

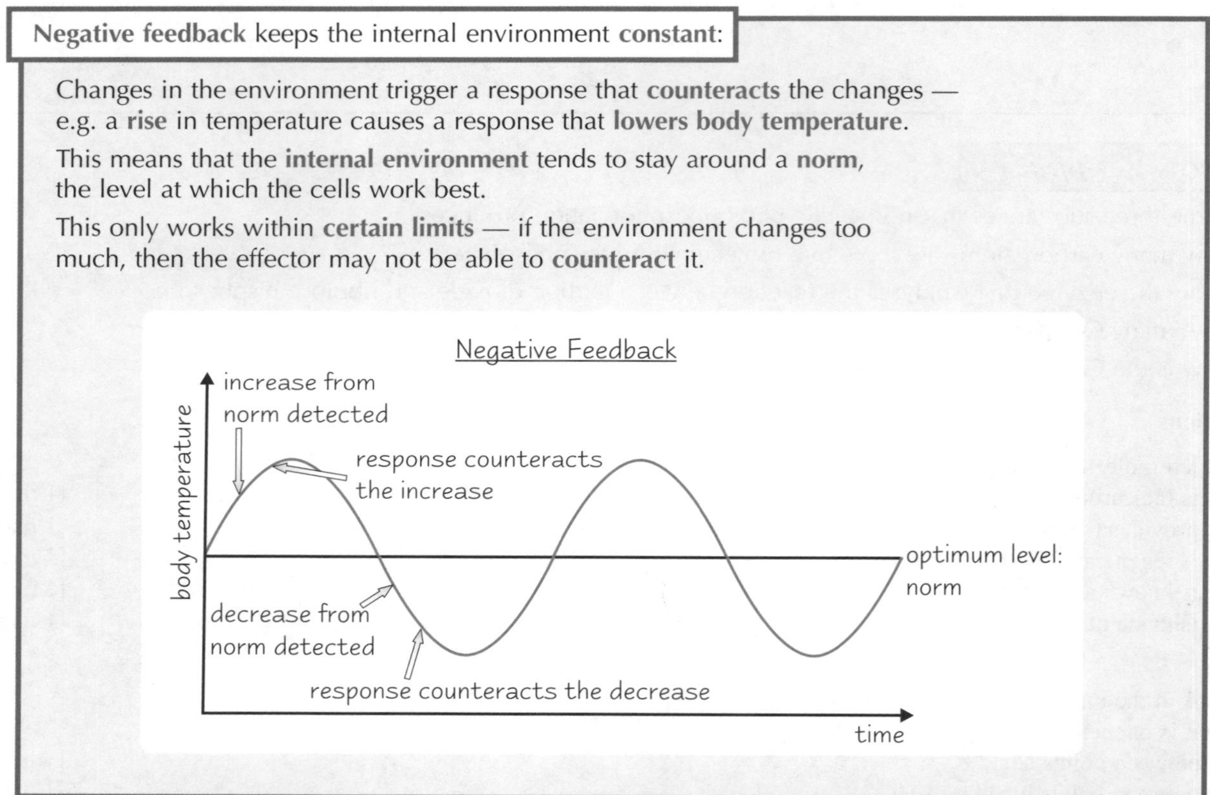

Mammals often use several **different responses** to keep a system in equilibrium.
This lets them control things better, e.g. temperature control and control of blood glucose concentration.

Homeostasis and Temperature Control

Mammals can Regulate their Body Temperature

The **skin** has a surface **epidermis**, and a thicker, deeper **dermis** with features for temperature control:

1) The skin has lots of **blood capillaries**. When you're too hot, the arterioles dilate (**vasodilation**), and more blood flows through the capillaries in the surface layers of the dermis to release more heat by **radiation**. When you're too cold, the arterioles constrict (**vasoconstriction**), reducing heat loss.
2) In mammals, **sweat** is secreted from **sweat glands** when the body is too hot. Sweat evaporates from the surface of the epidermis, using **body heat** and so cooling the skin.
3) Mammals have a layer of hair to provide **insulation** by trapping air, which is a poor heat-conductor. When it's cold, the **erector pili muscles** contract, which raises the hairs, trapping more air and preventing heat loss.

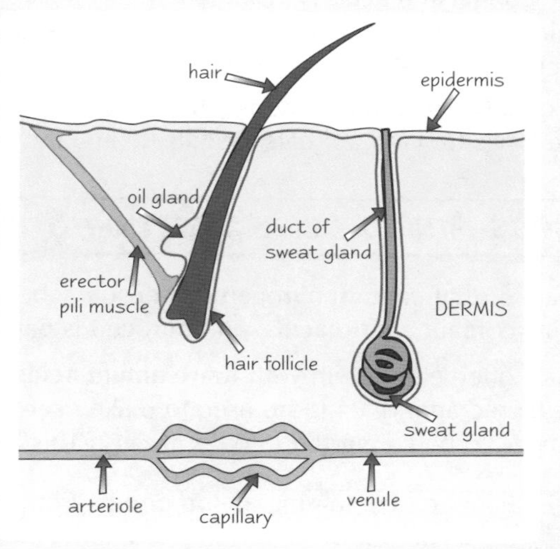

Thermoregulation is controlled by the Autonomic Nervous System

Body temperature is regulated by the unconscious actions of the **autonomic nervous system** (see pages 30-31):

1) If the external temperature **rises**, this stimulates thermoreceptors in the skin dermis, which send action potentials along **sensory neurones** to the hypothalamus. The hypothalamus sends **action potentials** along **motor neurones** to effectors in the skin. The effectors cause **vasodilation** of arterioles and more sweat secretion, so **more heat** is lost from the skin. The production of certain **hormones** is **decreased**, which decreases the metabolic rate so less heat is generated.

2) If the external temperature **falls**, the hypothalamus causes dermal arterioles to constrict and erector pili muscles to contract, erecting the hairs and trapping more air. The hypothalamus stimulates the production of a hormone to **increase** the **metabolic rate**, generating heat from **increased respiration**.

Practice Questions

Q1 Define homeostasis.

Q2 Give two factors that are controlled by homeostasis in the body of a mammal.

Q3 For one of your chosen factors, explain why it is beneficial to control it.

Q4 Name two ways in which the body cools itself down.

Q5 How does the hypothalamus control body temperature?

Exam Question

Q1 a) Explain what is meant by the term 'negative feedback'. [2 marks]

b) Give two examples of factors that are controlled by a negative feedback mechanism. [2 marks]

My biology teacher often gave me negative feedback...

The key to understanding homeostasis is getting your head round negative feedback. It's not complicated — if one thing goes up, the body responds to bring it back down, or vice versa. Look at pages 20-21 for more negative feedback loops.

Removal of Metabolic Waste

Here's how the body gets rid of things it doesn't need, like nitrogenous waste, carbon dioxide and Westlife albums.

Chemical Reactions in cells produce Waste Products

All the chemical reactions inside your cells make up your **metabolism**. Many of these chemical reactions produce substances not needed by the cell — some are even poisonous. These **waste** products need to be removed from the body by **excretion**.

For example, carbon dioxide is a toxic **waste product** of **respiration**. It is removed from the body by the **lungs** (in land animals) or **gills** (in aquatic ones). The lungs and gills act as **excretory organs**.

Excess Amino Acids Can't be Stored in the Body

Substances that contain nitrogen can't usually be stored by the body for later use. Proteins contain **amino acids**, and nitrogen is part of the **amino group**, NH_2.

Animals often eat protein with **more amino acids** than the body can use at once. Some amino acids are used to make **useful proteins**, but the **excess** ones need to be **converted** to other things. This happens in the **liver**:

1) Nitrogen-containing **amino groups** from the excess amino acids are removed, forming **ammonia** and **organic acids**. This is **deamination**. The organic acids are respired or converted to carbohydrate and stored as glycogen.
2) Ammonia is too poisonous for mammals to excrete it directly, so the **ammonia** reacts with **carbon dioxide** to form safer **urea**.
3) The urea is released into the blood, then **excreted** from the body by the **kidneys**.

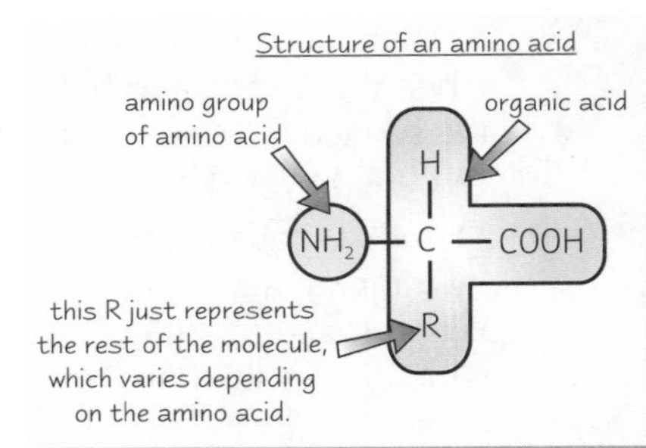

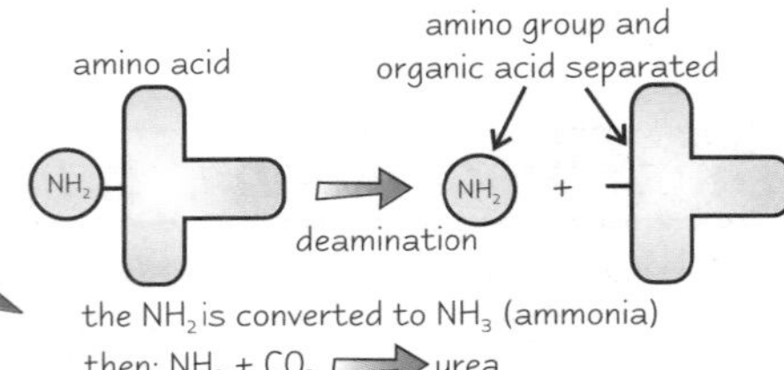

The Kidneys are Organs of Excretion

1) Urea produced by the liver is **excreted** from the body by the **kidneys**. Urea is dissolved in the blood plasma.
2) When the blood passes through the kidney nephrons (the tubes that run through the kidneys — see below for more on these), liquid is filtered out of the blood, carrying small solutes with it, including **urea**.
3) The useful solutes are reabsorbed, and the waste products are removed from the body in **urine**.

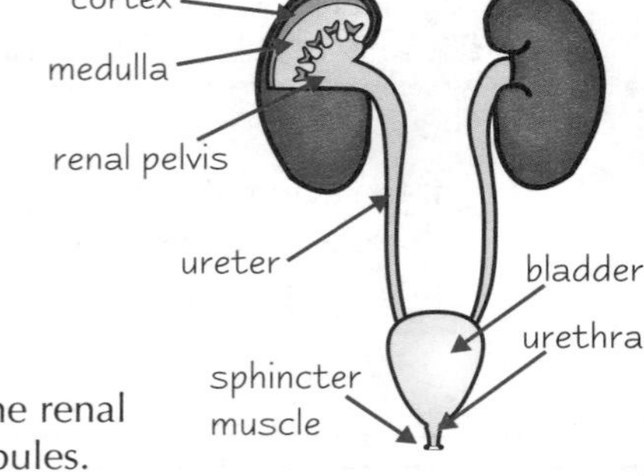

You need to know about **two main stages** of kidney function — **ultrafiltration** in the renal capsules (also called the Bowman's capsules) and **selective reabsorption** in the tubules.

Ultrafiltration in the kidneys takes place in the Renal Capsules

Blood enters the kidney cortex through the **renal artery**, then goes through millions of knots of capillaries in the kidney cortex. Each knot, (**glomerulus**) is a bundle of capillaries looped inside a hollow ball called a **renal** (or Bowman's) **capsule**. An **afferent arteriole** takes blood into each glomerulus, and an **efferent arteriole** takes blood out.

Ultrafiltration takes place in all body tissues that have capillaries. Blood pressure squeezes liquid from the blood through the capillary wall. Small molecules and ions pass through, but larger ones like proteins and blood cells stay behind in the blood. In most parts of the body, this liquid gathers between the cells as tissue fluid. In the kidney, the liquid collects in microscopic **tubules**. Useful substances are **reabsorbed** back into the blood, and waste substances (like dissolved urea) are excreted in the urine.

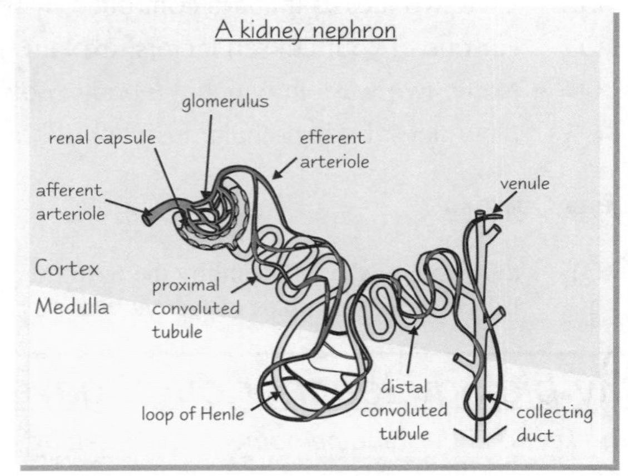

Removal of Metabolic Waste

Ultrafiltration in the capillary and renal capsule membranes

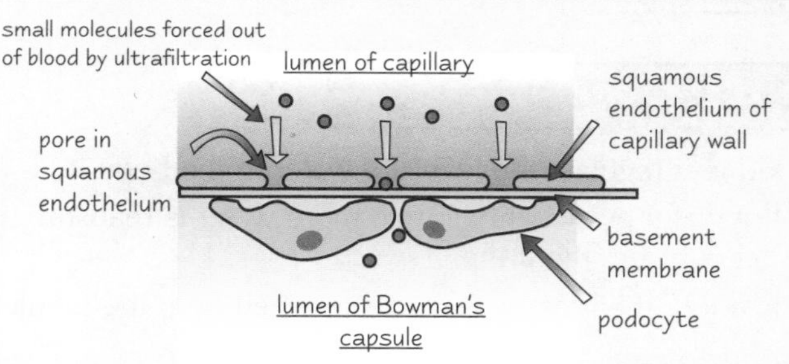

Small molecules and ions pass from blood in the glomerular capillary into the renal capsule through:

1) Pores in the **capillary wall** (the wall is made of one layer of flat cells — squamous endothelium).
2) A **basement membrane** made up of collagen fibres and glycoprotein.
3) A specialised epithelium of the renal capsule, made up of cells called **podocytes**. These support the membrane while letting the filtrate pass through.

Useful substances are *Reabsorbed*

Filtrate from the renal capsule enters the **proximal convoluted tubule** in the cortex of the kidney. The wall of this tubule is made of **cuboidal epithelium**, with **microvilli** facing the filtrate to increase the surface area.

Blood leaving the glomerulus along the efferent arteriole enters another capillary network that's wrapped around the proximal convoluted tubule. This provides a big surface area for **reabsorption** of useful materials from the **filtrate** (in the tubules) into the **blood** (in the capillaries) by:

1) **active transport** of glucose, amino acids, vitamins and some salts
2) **osmosis** of water

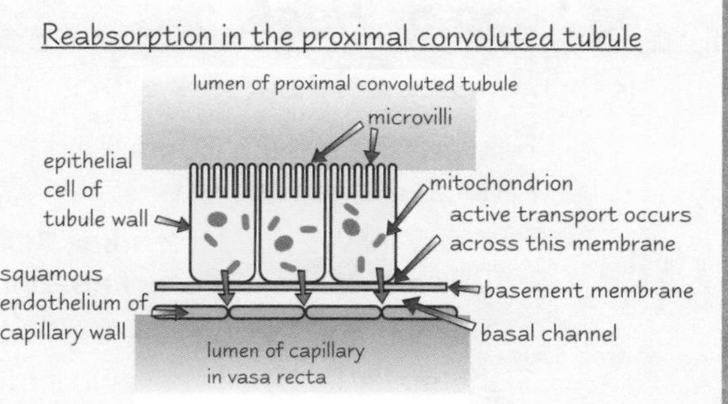

Water enters the blood by **osmosis** because the water potential of the blood is **lower** than that of the filtrate. Most of the water is reabsorbed from the **proximal convoluted tubule**. Reabsorption from the **distal convoluted tubule** and **collecting duct** is controlled by **hormones**.

Urine is a *Mixture* of substances dissolved in *Water*

Urine usually contains the following things:

1) **Variable amounts** of **water** and **dissolved salts**, depending on how much you've drunk.
2) **Variable amounts** of **dissolved urea**, depending upon how much protein you've eaten.
3) Other substances such as hormones and water-soluble vitamins.

It **doesn't** usually contain:

1) **Proteins**, which are too big to be filtered out in the renal capsule (they can't pass through the basement membrane).
2) **Glucose, amino acids or vitamins** — they're reabsorbed back into the blood from the proximal convoluted tubule.
3) **Blood cells.**

Practice Questions

Q1 What happens to excess amino acids that can't be stored in the body?

Q2 Name three components of the filtrate of a nephron that will be reabsorbed back into the blood from the proximal convoluted tubule.

Q3 What are the names of the processes that account for the reabsorption of substances?

Exam Question

Q1 a) Suggest how the features of the proximal convoluted tubule of a kidney nephron maximise the rate of absorption of glucose. [5 marks]

b) Suggest why a smaller quantity of urea passes from the tubule into the blood than glucose. [3 marks]

It's steak and excretion organ pie for dinner…

The kidneys are pretty complicated organs. That's why it's so serious when they go wrong — all that toxic urea would just stay in your blood and poison you. If your kidneys fail you'll end up hooked up to a machine for hours every week so it can filter your blood, unless some kind person donates a new kidney for you.

Removal of Metabolic Waste

You should have a rough idea about how the kidneys work by now, so here's a bit more detail on how they help with homeostasis. Read. Learn. Enjoy.

The **Kidneys** regulate the body's **Water Content (Osmoregulation)**

Mammals excrete urea in solution, so **water** is lost too. The kidneys regulate the levels of water in the body.

1) If the body is **dehydrated** (e.g. if the body has lost a lot of water by sweating), then more water is **reabsorbed** by osmosis from the tubules of the nephron, so less water is lost in the urine.
2) If the body has a **high water content** (e.g. from drinking a lot), then **less** water is reabsorbed from the tubules, so more water is lost in the urine.

This regulation takes place in the middle and last parts of the nephron — the **loop of Henle**, the **distal convoluted tubule** and the **collecting duct**. The **volume** of water reabsorbed is controlled by hormones.

The **Loop of Henle** has a **Countercurrent Multiplier Mechanism**

This sounds really scary, but stick with it, it's not too complicated really:

Just between the proximal and distal convoluted tubules is a part of the nephron called the **loop of Henle**. It's made up of two 'limbs'.

1) Water leaves the **descending limb** of the loop by osmosis, and Na^+ and Cl^- ions diffuse into the loop. This makes the **ion concentration** of the tubule **higher** towards the **base** of the loop.
2) The Na^+ and Cl^- ions are then actively pumped out of the top of the **ascending limb** of the loop into the medulla. The high concentration of Na^+ and Cl^- ions in the medulla causes water to leave the collecting duct and descending limb by **osmosis**.
3) This creates a **concentration gradient** within the kidney.

This mechanism is called the **countercurrent multiplier**.

The countercurrent multiplier mechanism lets land-living mammals produce urine with a **solute concentration** higher than that of the blood, so they can avoid losing too much water. The volume of water reabsorbed can be **regulated** depending on the needs of the body.

Another hard day reading, writing and countercurrent multiplying.

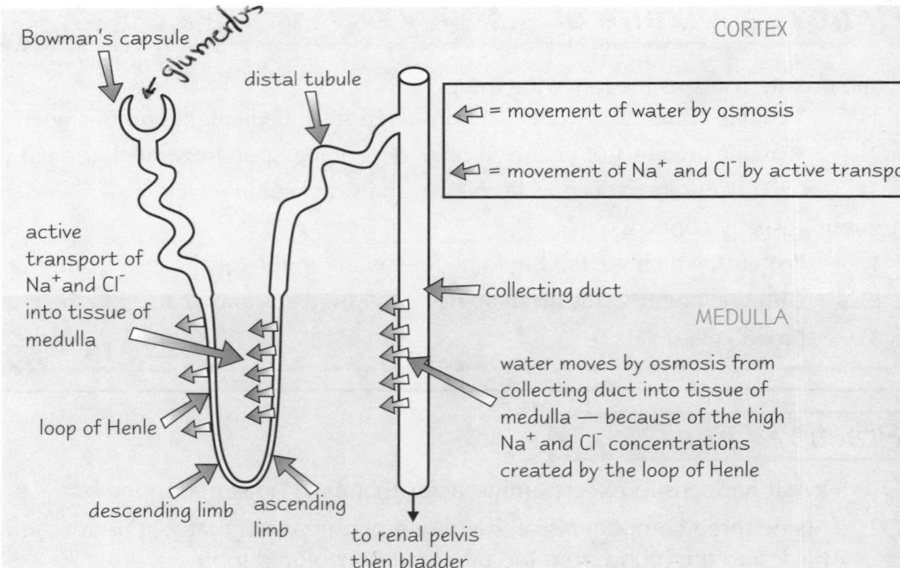

Water Reabsorption is controlled by **Hormones**

A hormone released from the **posterior pituitary gland**, called **antidiuretic hormone (ADH)**, makes the walls of the distal convoluted tubules and collecting ducts **more permeable** to **water**. More water is then **reabsorbed** from these tubules into the medulla (and then into the blood) by **osmosis**. This means that **less water** is lost in the urine.

It's called antidiuretic hormone because diuresis is when lots of dilute urine is produced.

Removal of Metabolic Waste

Blood ADH levels are **High** when you're **Dehydrated**

Dehydration is what happens when you **lose water**, e.g. by sweating during exercise:

1) The water content of the blood drops, so its **water potential drops**.
2) This is detected by **osmoreceptors** in the **hypothalamus**.
3) This stimulates the **pituitary gland** to release **more ADH** into the blood.
4) The ADH **increases** the **permeability** of the walls of the **collecting ducts** in the kidneys, so **more water** is **reabsorbed** back into the blood by **osmosis**.
5) **Less water** is lost in the urine.

Blood ADH levels are **Low** when your body is **Hydrated**

If you drink lots of water, more water is absorbed from the gut into the **blood**, and the **excess** is lost in the **urine**:

1) The water content of the blood rises, so its **water potential rises**.
2) This is detected by the **osmoreceptors** in the **hypothalamus**.
3) This causes the **pituitary gland** to release **less ADH** into the blood.
4) Less ADH means that the collecting ducts are less permeable, so **less water** is **reabsorbed** into the blood by **osmosis**.
5) **More water** is lost in the urine.

Desert Animals need to **Conserve Water**

Animals in a **hot**, **dry** environment have to control their water loss to survive, so they often have **special adaptations** (see the section on kangaroo rats on page 66 for more on this).

Desert **mammals** have very long loops of Henle, so there's a greater **surface area** to accumulate **more sodium chloride** ions in the medulla. This means their medullas have especially **low water potentials** so **more** water can be reabsorbed from the collecting ducts, making their urine more concentrated.

Practice Questions

Q1 What are the two main ways in which water can be lost from the body of a terrestrial animal?

Q2 Describe how the loop of Henle increases the salt concentration of the medulla of the kidney.

Q3 Explain the importance of the medulla in allowing reabsorption of water from urine.

Q4 Which gland releases ADH?

Q5 What is the effect of ADH on kidney function?

Exam Questions

Q1 Levels of ADH in the blood rise during strenuous exercise.
Explain the cause of the increase and the effects it has on kidney function. [10 marks]

Q2 Suggest why mammals adapted to life in dry deserts have longer loops of Henle. [5 marks]

If you don't understand what ADH does, urine trouble...

Seriously, though, there are two main things to learn from these pages — the countercurrent multiplier mechanism and the role of ADH in controlling the water content of urine. You'll need to be able to identify the different parts of the kidney nephron too. Keep writing it down until you've got it sorted in your head, and you'll be just fine.

Stimulus and Response

The nervous and endocrine systems are really important to animals. Plants don't have these like us, and that's why they tend to sit around doing nothing all day. These systems let us respond to our environment and tell what's going on.

Organisms have **Receptors** that are sensitive to **Stimuli**

A **stimulus** is any change in the environment that brings about a **response** in an organism — for example, a vibration that's detected by receptors in an organism's ears as a sound, or light that's detected by receptors in the organism's eyes as an image. A **receptor** is the part of the body of an organism that **detects** the stimulus.

You can classify receptors in animals depending on the **type of stimulus** they detect:

1) **Thermoreceptors** are sensitive to **temperature** — they're stimulated by **heat energy**.
2) **Photoreceptors** are sensitive to **light** — they're stimulated by **electromagnetic energy**, e.g. the cells that contain the pigments in the retina of the eye.
3) **Mechanoreceptors** are sensitive to **sound**, **touch**, **pressure** or even **gravity** — they're stimulated by **kinetic energy**.
4) **Chemoreceptors** are sensitive to **chemicals** — they're stimulated by **chemical energy**. They're involved in the senses of smell and taste.

Some kinds of chemicals in plants, such as **phytochromes**, are receptors — they detect **light**.

Receptor cells have **Excitable Membranes**

1) Receptor cells are **excitable**. This means that in their resting (unstimulated) state, their cell membranes have a **potential difference** across them — i.e. the receptor cells have a difference in **charge** across their cell membranes. There's a **negative** charge on the **inside** of the membrane, and a **positive** charge on the **outside**. This is generated by a combination of protein **ion pumps** and **channels** (see page 24).
2) When the receptor cell is stimulated, changes inside the cell affect the charge across the cell membrane. **Ions** (charged atoms) move into or out of the cell and alter the **charge** on each face of the membrane. The charges are **reversed**, creating a **generator potential**. The larger the stimulus, the larger the generator potential.
3) When the receptor cell is **excited** like this, it can transmit a signal to an **effector**, as long as the **generator potential** is big enough. Stimulated receptors that set up **nerve impulses** in nerve cells are called **transducers**.
4) The **minimum** size of stimulus needed to transmit a signal is called the **threshold stimulus**. Some kinds of receptor cells need a bigger stimulus than others to get a response (they have a **higher threshold**), so they're **less sensitive**.

The Receptors Communicate with **Effectors**

An **effector** is a part of the body that brings about a **response** to the signal from a receptor.
In animals, effectors are usually **muscles** (the response is **contraction**), or **glands** (the response is **secretion**).

There are two main ways that a receptor **communicates** with an effector:

1) The receptor produces a **chemical** (hormone) which binds to the effector. The chemical moves from receptor to effector by **diffusion** for short distances, or **mass flow** (transport in bulk) for long distances.
2) The receptor triggers a **nerve** (electrical) **impulse** in nerve cells, which stimulates the effector to respond.

It takes longer for a **chemical signal** to reach the effector than it does for an **electrical signal**. Unlike plants, animals have **nervous systems** (electrical) as well as hormone systems, so they can respond more **quickly** to changes than plants can.

Communication by electrical signalling

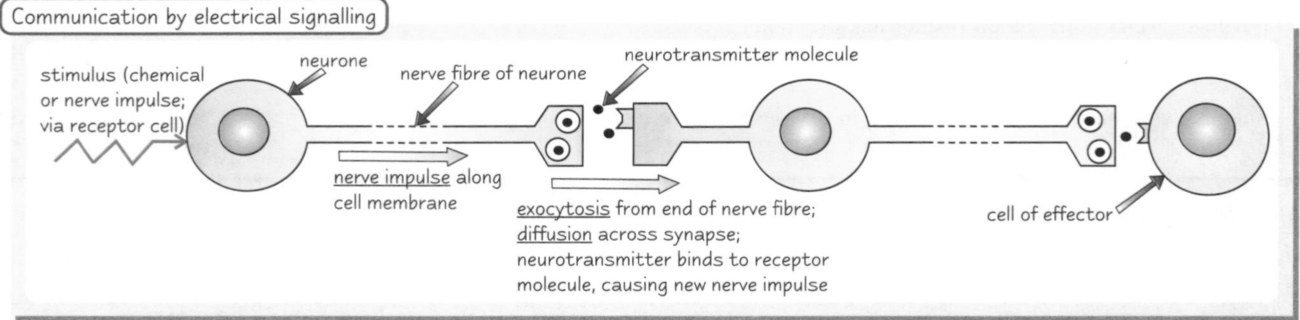

Stimulus and Response

The Endocrine System secretes Hormones

The endocrine system is made up of endocrine glands that secrete chemicals called hormones. A **gland** is any structure that is specialised for the **secretion** of one or more types of substance. Endocrine glands secrete **hormones directly** into the **blood** without using ducts. Many types of hormones are **proteins** or smaller **peptides**, e.g. **insulin** and **adrenaline**. Other hormones are fatty **steroids**, e.g. **oestrogen**, **progesterone** and **testosterone**.

1) Hormones are secreted when an **endocrine gland** is **stimulated**. Some glands are stimulated by a change in concentration of a specific solute (sometimes another type of hormone). Others are stimulated by **action potentials** arriving from the nervous system.
2) The hormone is **secreted** and **diffuses** into **blood capillaries**.
3) The hormone is circulated around the body by **mass flow** through the **bloodstream**.
4) The hormone diffuses out of the blood capillaries at different parts of the body. However, it will only bind to **cell-surface receptors** with complementary-shaped **binding sites**. This means that only **target cells** with the correct receptors will respond. Very **small** concentrations of hormones are needed to give a response in target cells. This response usually activates **enzymes** inside the cells.

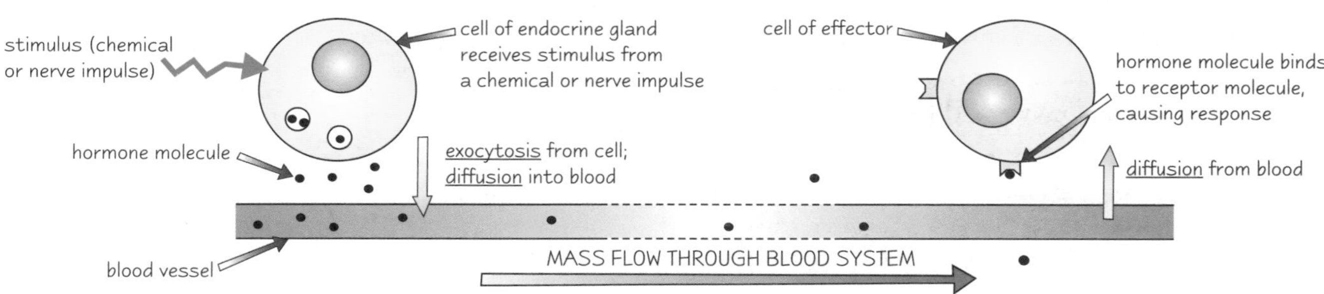

Hormonal Control is often by Negative Feedback

Hormones that **regulate** factors in the body, such as **glucose concentration,** often work by **negative feedback** (see pages 20-21). This is when a change in a factor is detected and the response **counteracts** the change.

Negative feedback control happens when hormones are working in **pairs**, e.g. insulin and glucagon (see the next pages). Here, an **increase** in blood glucose concentration stimulates the secretion of **insulin** to bring the glucose concentration back down. Then when blood glucose concentration **drops**, this stimulates secretion of **glucagon** to bring it back up. These pairs of hormones are said to work **antagonistically**.

Practice Questions

Q1 Why is it important that organisms can respond to stimuli in their environment?

Q2 What kinds of energy stimulate mechanoreceptors and photoreceptors?

Q3 What do receptors and effectors do in the nervous system?

Q4 Name one hormone which is a protein, and one which is a fatty steroid.

Exam Questions

Q1 a) Outline the two main ways in which the receptors of the nervous system communicate with the effectors. [3 marks]

b) Explain why the absence of one of these systems means that plants can respond only slowly to changes in their environment. [2 marks]

Q2 Chemoreceptors detect chemical stimuli. Suggest why exposure of a chemoreceptor cell to various different kinds of chemicals results in the generation of a nervous impulse for some chemicals, but not for others. [4 marks]

This page isn't as stimulating as I expected...

You should have some idea now about what receptors do and why they're important. It's basically so that organisms can respond to their environment. Being able to see, hear, feel, smell and taste allows them to find food and shelter, escape predators, avoid injury and illness — survive, really. The nitty gritty of how organisms respond is covered a bit later.

Regulation of Blood Glucose

Ever felt a 'sugar rush' after eating eighteen packets of Refreshers? OK, maybe no one else eats that many at once. These two pages are about how your body deals with all the glucose you get from your food.

Many Factors *can change the* ***Blood Glucose Concentration***

Glucose enters the blood from the **small intestine** and dissolves in the blood plasma.
Its concentration usually stays around **100mg per 100cm³ of blood**, but some factors can change this level:

1) Blood glucose concentration **increases** after consuming food, especially if it's **high** in carbohydrate.
2) Blood glucose concentration **falls** after exercise, because more glucose is used in respiration to release energy.

Big changes in the concentration of blood glucose can **damage cells** by changing the water potential.

The ***Pancreas*** *secretes* ***Hormones*** *to control* ***Blood Glucose Levels***

Two hormones, **insulin** and **glucagon**, regulate blood glucose concentration. They're secreted by clusters of cells in the **pancreas** called the **Islets of Langerhans**. These cells detect changes in blood glucose concentration.

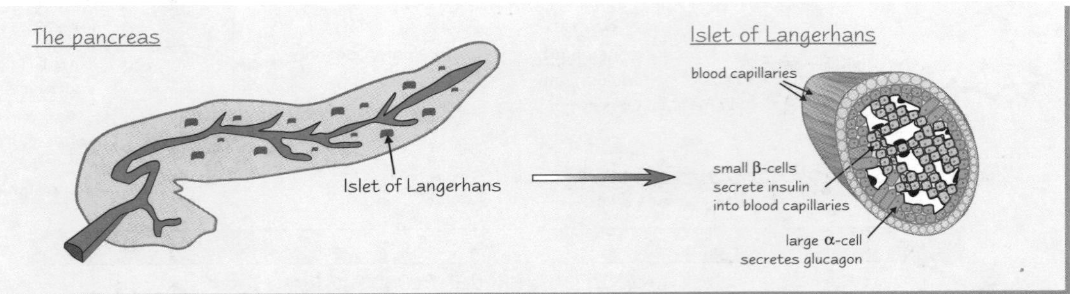

Controlling ***Blood Glucose Levels*** *is an example of* ***Negative Feedback***

Insulin and glucagon work together to regulate the blood glucose concentration by **negative feedback**:

When there's a rise in blood glucose concentration...

1) The glucose molecules bind to receptors in the cell membranes of small **beta (β) cells** in the Islets of Langerhans.
2) These cells secrete **insulin** into the blood.
3) Insulin molecules bind to receptors in the cell membranes of **hepatocytes** (liver cells) and other cells, e.g. muscle cells.
4) This **increases** the **permeability** of the hepatocyte cell membranes to glucose, so more glucose is absorbed.
5) Inside the hepatocytes, the insulin activates an **enzyme** that catalyses the condensation of glucose molecules into **glycogen**, which is stored in the cytoplasm of the hepatocytes, and in the muscles. This process is called **glycogenesis**.
6) Insulin also increases the rate of respiration of glucose in other cells. The blood glucose concentration **decreases**.

When there's a fall in blood glucose concentration...

1) The larger **alpha (α) cells** of the Islets of Langerhans secrete the hormone **glucagon** into the blood.
2) Glucagon binds to receptors on the **hepatocytes**.
3) This activates an **enzyme** inside the hepatocytes that catalyses the **hydrolysis** of stored glycogen into **glucose**. This process is called **glycogenolysis**.
4) The blood glucose concentration **increases**.

Regulation of Blood Glucose

Learn this Lovely **Negative Feedback** Diagram

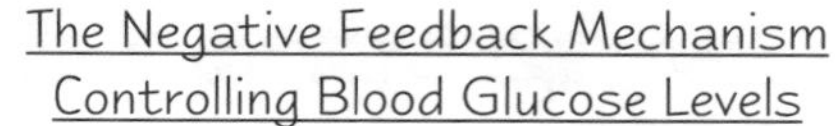

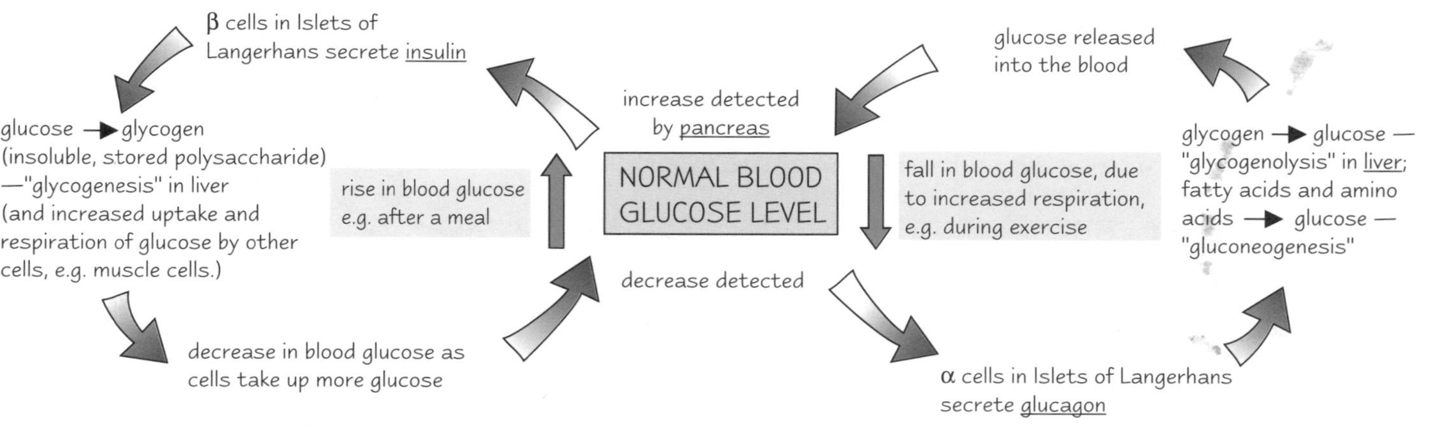

Other Hormones can affect **Blood Glucose Level** too

Other hormones can affect glucose levels in similar ways, although they bind to different receptors. For example, **adrenaline** is secreted during exercise (during a 'fight or flight response') and activates enzymes that **hydrolyse** stored glycogen into glucose, ready for increased muscle activity. It also increases the rate of respiration taking place inside the cells.

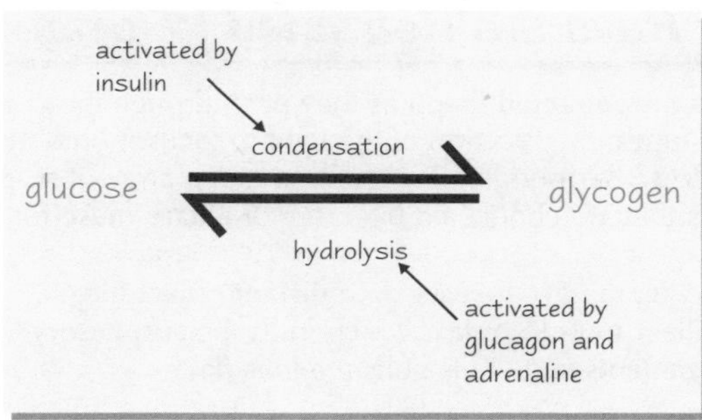

Practice Questions

Q1 Give one factor that will increase blood glucose concentration, and one that will reduce it.

Q2 What are the roles of the alpha and beta cells of the Islets of Langerhans?

Q3 State two effects of insulin on liver cells.

Q4 What is glycogenesis?

Q5 What is glycogenolysis?

Q6 Name two hormones that increase the blood glucose concentration.

Exam Questions

Q1 Explain the role of the endocrine system in returning the blood glucose level to normal after consuming a meal that has a high glucose content. [10 marks]

Q2 Explain why glucagon levels in the blood increase during exercise. [5 marks]

My glucose levels are low — pass the chocolate...

When you saw the words 'negative feedback' on these pages, did your heart just sink? Negative feedback is probably the most boring, and yet one of the trickiest concepts that you have to know about. But once you've got your head round it, you're laughing. Now eat a huge meal, wait for your blood glucose levels to rise, then draw that diagram until you know it.

The Mammalian Eye

For some reason the eye seems more interesting than most of the other subjects in this section. Maybe it's because we're very aware of our own eyes — more so than our muscles or our nerves, anyway.

The Eye is an organ with Photoreceptors for detecting light

1) The mammalian eye is a fluid-filled ball bound by a tough external **sclera**, which forms the transparent **cornea** in front.
2) A thinner transparent **conjunctiva** covers the cornea.
3) The inner lining of the back of the eyeball is the photoreceptive **retina**.
4) Between the sclera and the retina is the **choroid**, which is a layer rich in blood vessels to supply the retina and covered with pigment cells to prevent internal reflection of light.
5) The shape of the eyeball is maintained by the hydrostatic pressure of the **aqueous humour** behind the cornea (a clear salt solution) and the jelly-like **vitreous humour** behind the lens.
6) Light rays pass through the **pupil** (hole in the front) and are focused by the **lens** onto the **fovea** of the retina.
7) Action potentials are then carried from the retina to the brain by the **optic nerve**, a bundle of sensory neurones.
8) The **blind spot** is where the optic nerve leaves the eye. There are no photoreceptors there so it's **not** light sensitive.

suspensory ligaments, lens, sclera, choroid, retina, conjuctiva, cornea, aqueous humour, pupil, iris, position of fovea, blind spot, ciliary muscle, vitreous humour, optic nerve

The **iris** is a muscular diaphragm surrounding the **pupil**. It controls the amount of **light** entering the eye. In bright light the **circular muscle** of the iris **contracts** and the **radial muscle relaxes**, making the pupil **smaller**. Less light enters the eye, preventing **damage** to the **retina**. The opposite happens in dim light — the iris makes the pupil **dilate** to allow more light in.

Light Rays are Refracted to focus on the Retina

Light rays are **refracted** (bent) as they pass through the **cornea** and the **lens**. Most refraction happens at the cornea, and then the lens **fine-tunes** the direction of the light to focus it onto the **retina**. It does this by changing its **shape** so that the light is refracted more or less. Surrounding the lens is a radial array of **suspensory ligaments** that are connected by a ring of **ciliary muscles**:

1) When the eye focuses on a **distant** object the ciliary muscles **relax**, which pulls the **suspensory ligaments** taut. This pulls the lens **flat**.
2) When the eye focuses on a **near** object the ciliary muscles **contract**, so the **suspensory ligaments** are slack. This gives the lens a more **rounded** shape.

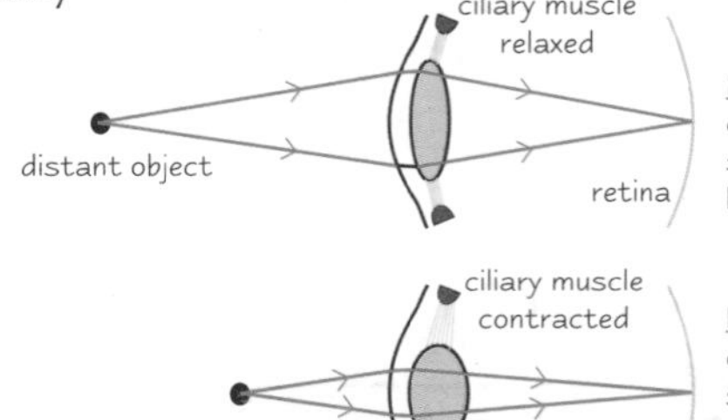

Light from a Distant Object
circular ciliary muscles relaxed; suspensory ligaments taut; lens pulled thin; light focused on retina.

Light from a Near Object
circular ciliary muscles contracted; suspensory ligaments slack; elastic lens more convex; light focused on retina.

The Retina is made up of Three Main Layers

1) **Photoreceptors** (**rods** and **cones**) which have **outer segments** and **inner segments**:
   - The outer segments have lots of **flattened vesicles** containing **pigments** that absorb light energy. When light is detected, there's a **chemical change** inside the cell which creates a **generator potential** on the cell membrane.
   - The **inner segment** of each receptor has **mitochondria** and a **nucleus**, and connects with a **synapse**.

   Rods and **cones** differ in shape (see diagram). There are **three types** of cones, each with a different type of **pigment**.
2) There's a layer of **bipolar neurones** (with some **cross connections**). The cross connections between the bipolar neurones allow the signals sent from the cells of the retina to be **coordinated**.
3) A layer of **sensory neurones** have axons leading to the **optic nerve**.

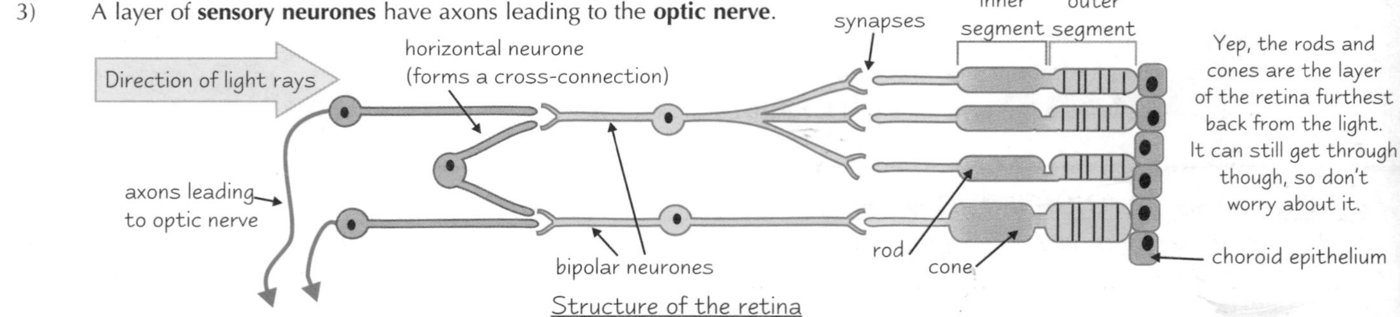

Structure of the retina

The Mammalian Eye

Outer Segments contain *Visual Pigments* that are *Proteins*

Rods use a visual pigment called **rhodopsin**. **Cones** use three forms of another pigment called **iodopsin**.
This is how **rhodopsin** is used to send impulses to the brain:

1) Light energy is absorbed by a part of the rhodopsin pigment called **retinal**. This changes shape from one form to another, and detaches from the protein part called **opsin**. This process is known as **bleaching**.
2) **Bleaching** causes the cell membrane of the **outer segment** to become **hyperpolarised**. This means that **sodium channels** close, so the **resting potential** value gets more negative (from about -65 mV to about -120 mV).
3) Less **inhibitory neurotransmitter** is released across the synapse, so there's less **inhibition** of the **bipolar neurone**. This means the cell membrane of the bipolar neurone becomes **depolarised**.
4) An **action potential** is formed in the bipolar neurone membrane, and is transmitted to the brain through the **optic nerve**.

Retinal and **opsin** then join back together using an enzyme-catalysed reaction. This **regenerates** the pigment so that it's ready to be stimulated again.
The same sort of thing happens with the **iodopsin** in **cones**, but it breaks apart **less easily** and joins back together **more slowly**. This means that cones are best suited to **higher light intensities**, and we depend more on rods in **dim light**.

Rods are more *Sensitive*, but *Cones* let you see more *Detail*

1) There are about **twenty times** more rods in the human eye than there are cones. The cones are mostly found packed together in the **fovea**, which is where most of the light that enters the eye tends to focus. They give better **visual acuity** (clarity) than rods do, and let us see images more **accurately** and in more **detail**. This is because each cone synapses with its **own individual bipolar synapse**, so it can send more detailed information to the brain. Cones also give **colour vision** (see below).

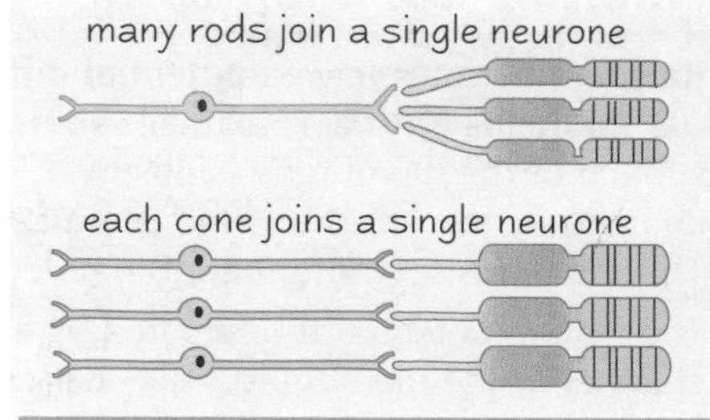

2) The rods are found outside the fovea, in the more peripheral parts of the retina. They're a lot more **sensitive** than cones, because lots of rods converge onto the **same bipolar neurone**. This means that even small responses from several rods can combine and send a message to the brain. But having lots of rods converging onto the same bipolar neurone also means that they can't provide as much **clarity** or **detail** as the cones can. This is why an object you can see 'out of the corner of your eye' isn't very clear. You have to move your eyes so that the light from the object focuses on your **fovea** to see it clearly.
3) Colour vision is explained by the **trichromatic theory**. It states that all the **colours** we see come from stimulation of the three types of **cone** in different proportions. Each type of cone is most sensitive to one of **red**, **blue** or **green** light. Any particular colour is then experienced because the wavelength stimulates one, two or all three of these types to a different degree.

Practice Questions

Q1 Give the functions of the following parts of the eye:
a) choroid b) ciliary muscle c) iris.

Q2 What shape is the lens when the eye focuses on a distant object?

Q3 What happens to rhodopsin when it absorbs light energy?

Q4 Which are more sensitive, rods or cones?

Q5 What three colours of light are the three types of cone most sensitive to?

Exam Question

Q1 Explain, with reference to rods and cones, how the human eye has both high sensitivity and high acuity. [8 marks]

Be thankful for the eye — without it, you wouldn't see this lovely page...

OK, so maybe that's not the best reason to be thankful for the eye. In fact, it might be enough to make you wish it had never evolved. But too late, it has, so there's no excuse not to learn about it. Those rod and cone diagrams are just to illustrate a point, but the diagram of the eye and those two about the lens have to be learnt. The best way is to practise drawing them yourself.

The Nerve Impulse

It's no good being aware of your environment if you can't do anything about it. That's where neurones come in.

Nerve Cells have Polarised Membranes so they can carry Electrical Signals

1) The **nervous system** is made up of nerve cells called **neurones**. Each neurone consists of a **cell body** and extending **nerve fibres**, which are very thin cylinders of cytoplasm bound by a cell membrane. Neurones carry waves of electrical activity called **action potentials** (nerve impulses). They can carry these impulses because their cell membranes are **polarised** (see below) — there are different **charges** on the inside and outside of the membrane.

2) The nerve fibres let the neurones carry action potentials over **long distances**. There are tiny gaps, called **synapses**, between the different nerve fibres. Action potentials can't cross, so a chemical called a **neurotransmitter** is secreted at the tip of each nerve fibre to cross the gap. This stimulates a **new** action potential in the next nerve fibre on the other side of the synapse (see pages 26-27).

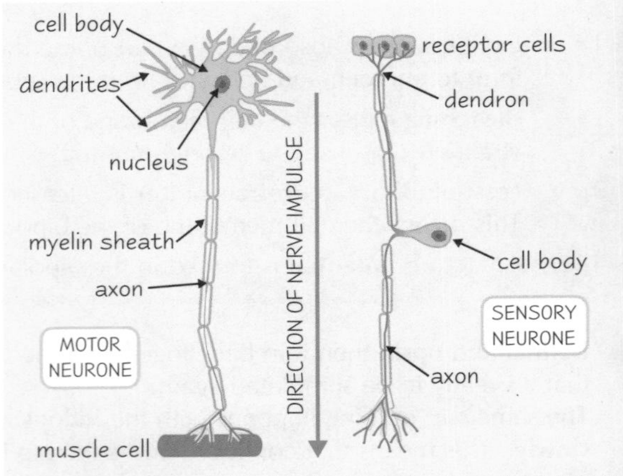

3) Between the receptors and the effectors, the **central nervous system** (i.e. brain and spinal cord) **coordinates** the action potentials passing through the nervous system. **Sensory neurones** carry action potentials from receptors to the central nervous system. **Relay neurones** carry action potentials through the central nervous system, and **motor (effector) neurones** carry them from the central nervous system to effectors.

Neurone cell membranes are Polarised when they're Resting

Resting neurones have a **potential difference** (a difference in **charge**) of about **-65 millivolts** (mV) across their cell membranes. This is because the **outer** surface of the membrane is **positively** charged and the **inner** surface **negatively** charged — the -65 mV is the **overall difference** in charge between them. This is the **resting potential** of the membrane, which is said to be **polarised**.

The resting potential is generated by a **sodium-potassium pump** and a **potassium channel** in the membrane. The sodium-potassium pump moves **three sodium ions** out of the cell by **active transport** for every **two potassium ions** it brings in. The potassium channel then allows **facilitated diffusion** of potassium ions back out of the cell. The outer surface of the membrane becomes more positive than the inner surface because overall, more positive ions move **out** of the cell than move **in**.

Sodium-potassium pump and potassium channel

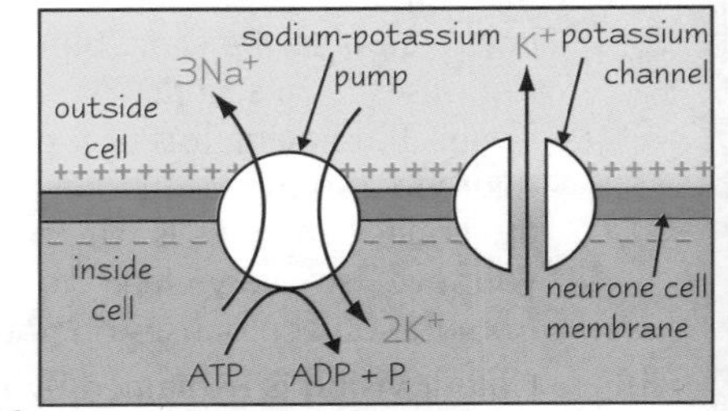

Neurone cells become Depolarised when they're Stimulated

The **sodium-potassium pumps** work pretty much all the time, but **channel proteins** (like the potassium channels) can be opened or closed. **Depolarisation** of neurone cell membranes involves another type of channel protein, **sodium channels**. If a neurone cell membrane is stimulated, sodium channels **open** and **sodium ions** diffuse in. This **increases** the positive charge **inside** the cell, so the charge across the membrane is **reversed**. The membrane now carries a potential difference of about **+40 mV**. This is the **action potential** and the membrane is **depolarised**.

When sodium ions diffuse into the cell, this stimulates nearby bits of membrane and **more** sodium channels open. Once they've opened, the channels automatically **recover** and close again.

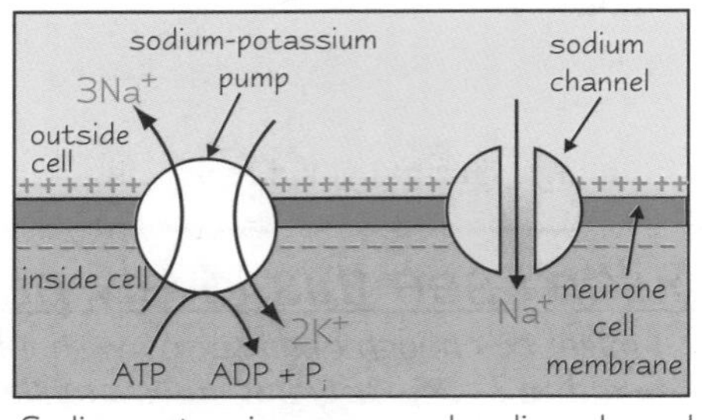

Sodium-potassium pump and sodium channel

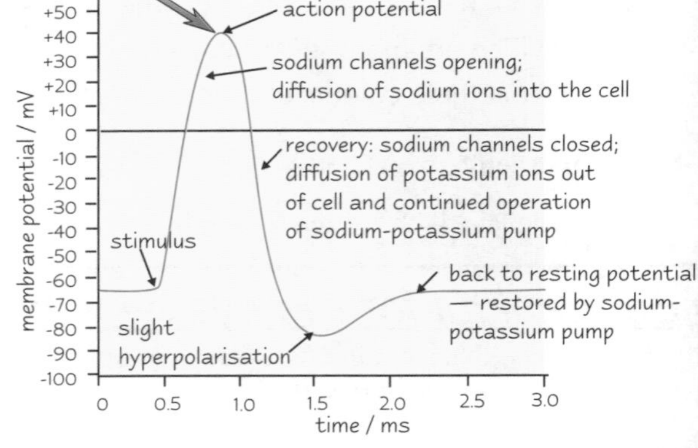

The Nerve Impulse

Remember these important features of *Action Potentials*

1) **Nerve axons** in vertebrates are usually covered in a layer of **myelin sheath**, which is produced by **Schwann cells**. Myelin is an **electrical insulator**. Between the sheaths there are tiny patches of bare membrane called **nodes of Ranvier**, where sodium channels are **concentrated**. Action potentials **jump** from one node to another, which lets them move **faster** (this is called **saltatory conduction**).

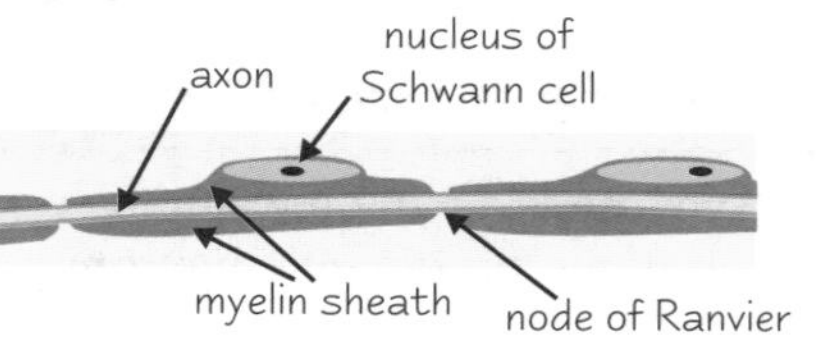

2) Action potentials also go faster along axons with **bigger diameters**, because there's less **electrical resistance**.

3) They go faster as **temperature** increases too, up to around **40°C**. After that, the proteins begin to **denature**.

4) Action potentials have an **all-or-nothing** nature. This means that the values of the resting and action potentials for a neurone are **constant**, and you can't get anything in between. (A **bigger stimulus** just increases the **frequency** of the action potentials. The **strength** of the action potentials stays the same.)

5) A **threshold stimulus** must be applied to get an action potential (see page 18).

6) Straight after an action potential has been generated, the membrane enters a short **refractory period** when it can't be stimulated, because the sodium channels are **recovering** and can't be opened. This makes the action potentials pass along as **separate signals**.

7) Action potentials are **unidirectional** — they can only pass in one direction. This is because of the refractory period — it stops the impulse going back on itself.

Practice Questions

Q1 What do sensory, relay and motor neurones do in the nervous system?

Q2 What happens to sodium ions when a neurone membrane is stimulated?

Q3 Give two factors that increase the speed of conduction of action potentials.

Q4 What is meant by the 'all-or-nothing' nature of action potentials?

Exam Questions

Q1 The graph shows an action potential 'spike' across an axon membrane following the application of a stimulus.

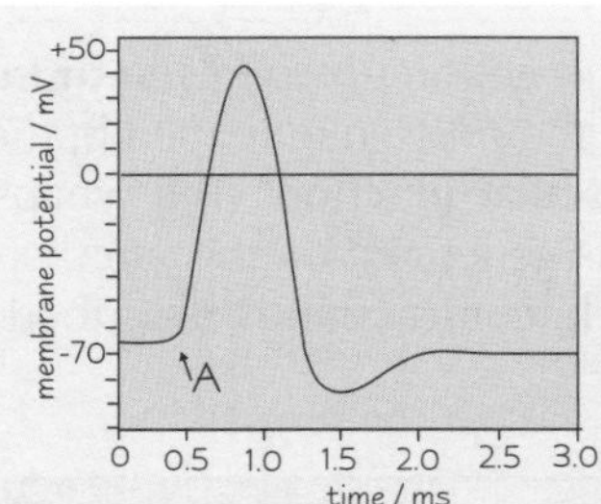

a) What label should be added at point A? [1 mark]

b) Explain what causes the change in potential difference from -65 to +40 mV. [3 marks]

c) Another stimulus was applied at 1.5 ms, but failed to produce an action potential. Suggest why. [2 marks]

Q2 Multiple sclerosis is a disease of the nervous system characterised by damaged myelin sheath. Suggest and explain how this will affect the transmission of action potentials. [5 marks]

I'm feeling a bit depolarised after all that...

The nervous system can seem like a really hard subject at first, but once you've gone over it a couple of times it starts to make sense. Nerves work because there's a charge across their membranes, and it's a change in this charge that sends the message along the nerve. The charge is set up using ions, which can then be pumped in and out to change the charge.

Synapses and Synaptic Transmission

This page is all about synapses, which are the little gaps between the end of one neurone and the start of the next one. Seems like quite an insignificant little thing to fill a whole two pages with, but never mind.

There are **Gaps** between **Neurones**

A **synapse** is a gap between the end of one **neurone** and the start of the next. An action potential arrives at the end of the axon of the **presynaptic neurone** (the neurone before the synapse), where there's a swelling called a **bouton** or **synaptic knob**. This has **vesicles** containing a chemical **neurotransmitter**, and the impulse passes across the synapse as follows:

1) The action potential opens **calcium channels** in the membrane, allowing calcium ions to diffuse **into** the bouton. Afterwards these are pumped back out using ATP.
2) The increased concentration of calcium ions in the bouton causes the **vesicles** containing the **neurotransmitter** to move up to and to fuse with the **presynaptic membrane**. This also requires ATP (it's an active process).
3) The vesicles **release** their neurotransmitter into the **synaptic cleft** (this is called **exocytosis** and it's an active process too).
4) The neurotransmitter **diffuses** across the synaptic cleft and binds to **receptors** on the **postsynaptic membrane** of the other neurone.
5) This stimulates an **action potential** in the postsynaptic membrane by opening the **sodium channels** (see page 24).
6) An **enzyme** is sometimes used to **hydrolyse** the neurotransmitter, so the response doesn't keep on happening. The neurotransmitter may also be taken back up into the presynaptic bouton, ready to be used again.

Because the receptors are only on the **postsynaptic** membranes, a signal can only pass across a synapse in **one direction** (it's **unidirectional**). The postsynaptic cell behaves as a **transducer**, because the **chemical** stimulus (neurotransmitter) is converted into an **electrical** one (action potential).

There are lots of **mitochondria** in the bouton of a neurone. These provide the **ATP** to make more neurotransmitter, power exocytosis and pump calcium ions out of the bouton.

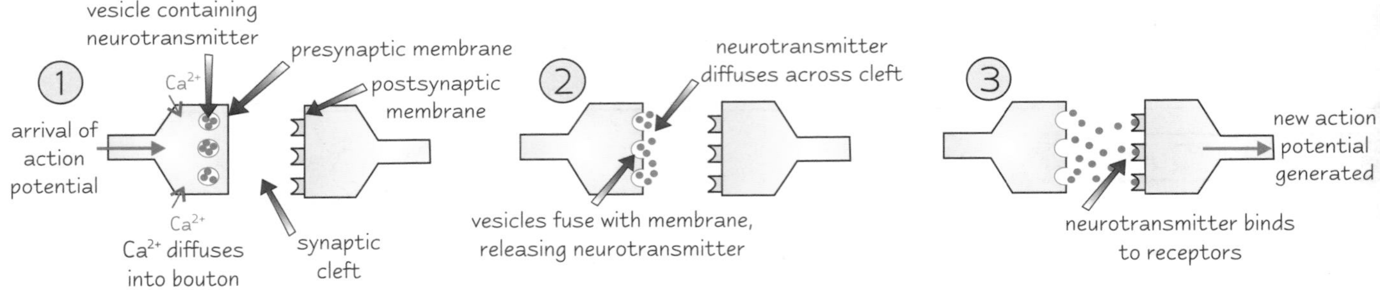

There's a **Synapse** between a Motor Neurone and Muscle Fibres

Motor neurones carry action potentials to **muscle fibres**. There's a synapse between the presynaptic membrane of the motor axon and the postsynaptic membrane of the muscle fibre (the **sarcolemma** — see page 32). This region is called the **neuromuscular junction**. The synapse functions in the same way as a synapse between two neurones, and an action potential is generated on the sarcolemma in the same way too. What happens afterwards to bring about **muscle contraction** is described on pages 32-33.

There are **Different Kinds** of **Neurotransmitters**

There are many kinds of neurotransmitters. A lot of neurotransmitters are **excitatory** (as described above), but some are **inhibitory**. These make the membrane **hyperpolarised**, creating a resting potential value that's even more **negative** than the usual -65 mV. This makes it harder to **excite**.

Some kinds of neurones in particular parts of the body only have receptors for certain **types** of neurotransmitter, e.g:

1) The neurotransmitter **acetylcholine** binds to **cholinergic receptors**. It's an excitatory neurotransmitter at many neurones and neuromuscular junctions. However, it **inhibits** contraction of **cardiac** and **smooth** muscle (but only because it excites inhibitory neurones).
2) **Noradrenaline** binds to **adrenergic receptors**, and it's generally **excitatory**.

Synapses and Synaptic Transmission

Some Drugs and Poisons affect the action of neurotransmitters

- **Agonists** are chemicals that **mimic** the effects of neurotransmitters. An example is **nicotine**, which binds to **cholinergic** receptors and stimulates **neuromuscular junctions**.
- **Antagonists block** the effects of neurotransmitters. An example is **curare**, a chemical that binds to **cholinergic receptors**, but blocks – rather than mimics – the acetylcholine molecules. This **paralyses** the muscle.

Reflex responses involve Reflex Arc Pathways

A **reflex** is an **involuntary stereotypical** response of part of an organism to an applied stimulus. This means that there's usually no **conscious control** over it, and it always produces the **same** kind of effect. It works because there are special patterns of neurones that make up **reflex arcs**. The simplest is a **monosynaptic reflex**, where the sensory neurone connects **directly** to the motor neurone so there's only **one synapse** within the central nervous system in the arc. An example is the **knee jerk reflex**. These reflex arcs often play a role in controlling **muscle tone** and **maintaining posture**. The response doesn't **require** action potentials to pass to the **brain**, so no conscious thought is needed for them to happen.

A **polysynaptic reflex** has at least **two synapses** within the central nervous system, due to the presence of a **relay neurone**. Action potentials **can** pass to the brain, so some conscious thought might be involved. An example of such a reflex is the quick response if you touch something hot.

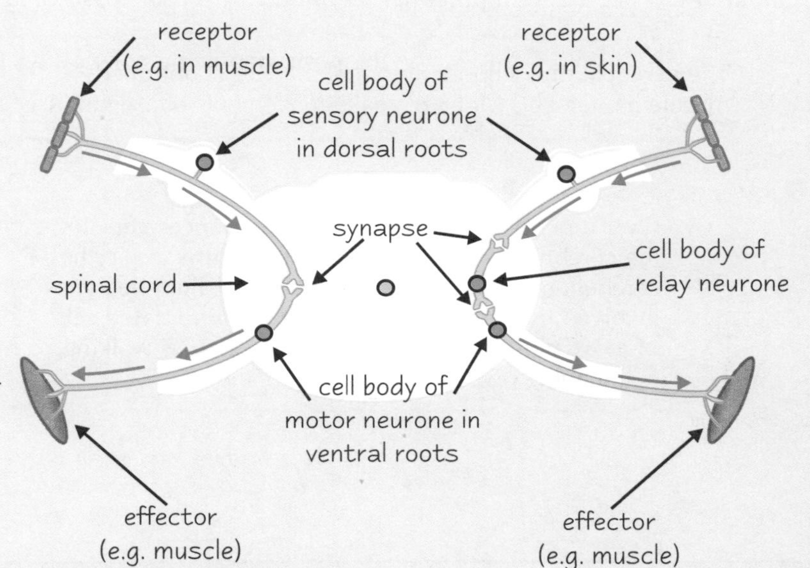

Ventral roots and dorsal roots are just the parts of spinal nerves that join with the spinal cord.

Transverse section through spinal cord showing monosynaptic reflex arc (left) and polysynaptic reflex arc (right)

Practice Questions

Q1 Why are there a lot of mitochondria in the bouton (synaptic knob) of the neurone?

Q2 Give two examples of a neurotransmitter.

Q3 What effect do antagonists have on neurotransmitters? Give an example of an antagonist.

Q4 What is a reflex response?

Q5 What's the difference between a monosynaptic reflex and a polysynaptic reflex?

Exam Questions

Q1 Describe the sequence of events leading from the arrival of an action potential at a bouton to the generation of a new action potential on a post-synaptic membrane. [8 marks]

Q2 Explain how the structure of the synapse ensures that signals can only pass through it in one direction. [4 marks]

Bouton — like button, but in a weird French accent...

Not the most exciting page in the book, but pleasantly dull I'd say. There's nothing too hard there. Some chemicals with annoyingly long names cross a gap so action potentials can move from neurone to neurone through the body. Reflex arcs usually don't have many synapses, so the response can happen quickly. Some, like the knee jerk reflex, only have one.

The Brain

And now ladies and gentlemen, the most important bit of all — the central nervous system, the controller of almost everything you do. It's like a giant computer, only much more complicated...

The **Central Nervous System** has **Two Parts**

The central nervous system is made up of your **spinal cord** and your **brain**:

1) The **spinal cord** runs through the protective, bony vertebral column.
The sensory and motor **spinal nerves** are connected to it.
Quick **reflex** reactions like sneezing can be processed by the spinal cord, without using the brain.
2) The **brain** is a massive bunch of **relay neurones** (see p. 24) connected to the spinal cord. The brain is inside in the protective bony **cranium** (skull), so the nerves connected to the brain are called **cranial nerves**.

The **Cerebellum** and **Cerebrum** control **Complex Conscious Behaviour**

They each have a highly folded **cortex** (the outer bit). They coordinate sensory input with motor output. Action potentials arriving along sensory nerve fibres pass along lots of possible pathways via the relay neurones, so each **input** has lots of **different potential responses**. This is the basis of **memory** and learning from experience.

The **cerebellum** is important for balance and controlled muscle movements. It helps make movements **coordinated**.

Over a period of time, **learning experiences** affecting the cerebrum influence the **neurone pathways** in the cerebellum. This means that the cerebellum becomes involved in controlling more **skilled coordinated sets of movements** that become routine (such as walking, maintaining posture or playing a piano).

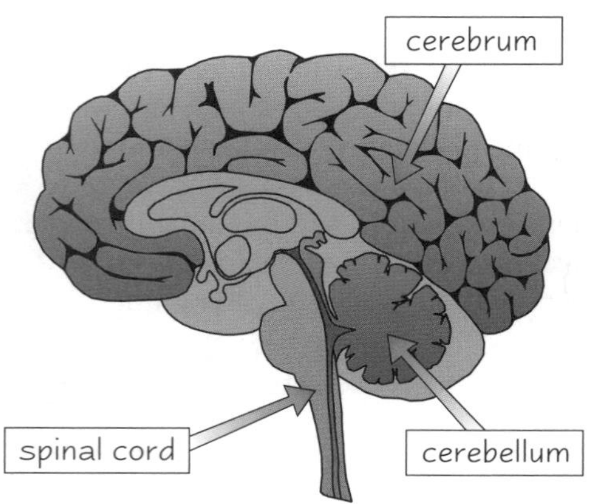

The **Cerebrum** is divided into **Two Cerebral Hemispheres**

The **cerebrum** is the largest part of the brain. It deals with the **voluntary activities** of the body. It's associated with advanced mental activity, like **emotion**, **memory** and **language**. It's divided into **two parts** (hemispheres). The **right** hemisphere deals with actions for the **left** side of the body, and vice versa.

1) Action potentials from particular parts of the body arrive in specific areas of the cerebral cortex (the outer bit) called **primary sensory areas**, e.g. impulses from the optic nerve of the eye go to the visual cortex, where they're interpreted by the brain.
2) Sensory areas pass action potentials to **association areas** where information is integrated.
3) **Motor areas** send impulses to effectors (muscles), causing movement.

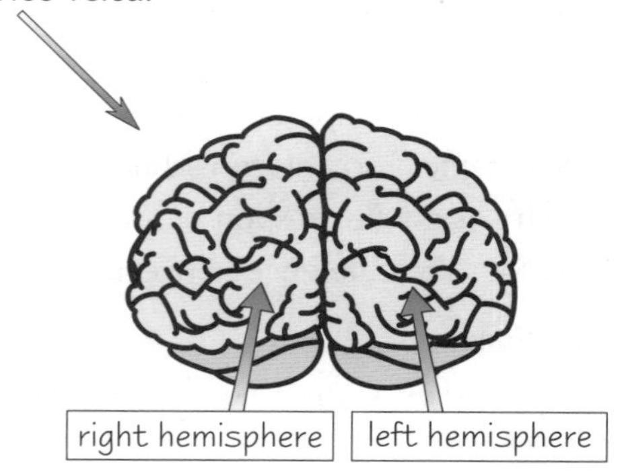

This brain is seen from the front. The left and right hemispheres are named as they're found in the body, so they look the wrong way round on diagrams. Just like with diagrams of the heart.

The Brain

Speech is controlled by the **Cerebrum**

1) **Receptors** in the ear send information to a **sensory area** of the cerebrum — the **primary auditory area**.
2) This sends information to the **auditory association area** (**Wernicke's area**), where words are identified, using the memory.
3) The information then goes to the **auditory motor area** (**Broca's area**) which coordinates muscle movement to produce speech for a response.
4) Speech involves **muscles** in the face, lips and tongue.

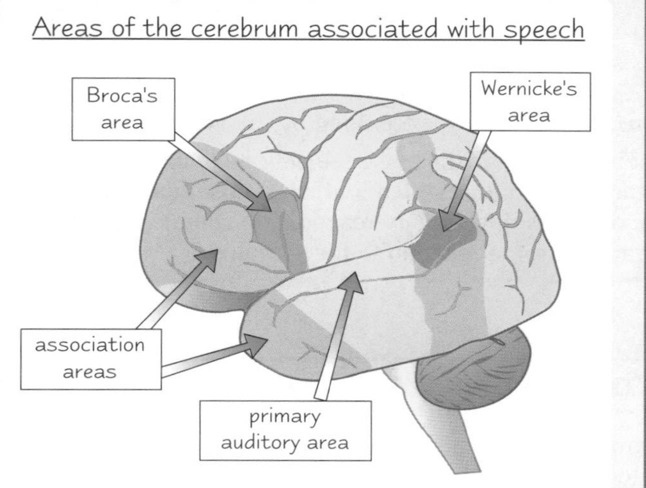

The **Size** of the **Controlling Areas** depends on the **Complexity** of **Innervation**

Some sensory, association and motor areas in the brain are bigger than others. The size of the area depends on the **complexity** of its **innervation** (in other words how big the input of information from sensory cells is).

The **visual cortex** is a very large **sensory area** located at the back of the cerebrum. It is large because it receives a very **large input** (complex innervation) from the **receptors** in the retina of the eye. The **visual association area** is also large because it is innervated by the visual cortex. This area interprets the sensory input.

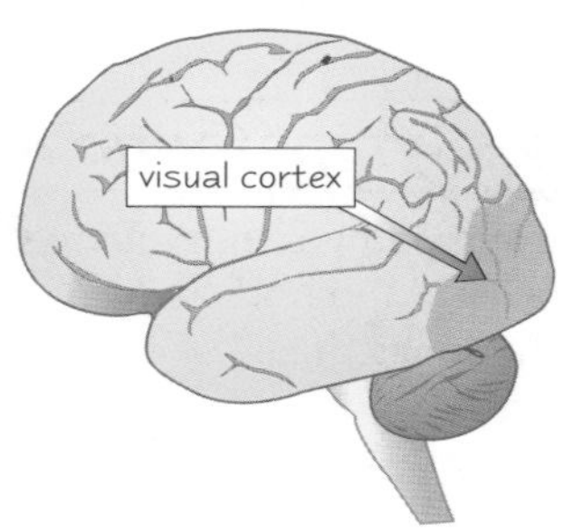

Practice Questions

Q1 Name the two main parts of the central nervous system.

Q2 Which side of the body does the right cerebral hemisphere control?

Q3 What is the auditory association area known as?

Q4 What is the auditory motor area also known as?

Q5 Which area in the brain interprets information from the visual cortex?

Exam Questions

Q1 Describe the functions of the sensory areas and the association areas of the brain's cerebrum. [8 marks]

Q2 A man cannot speak after suffering a head injury, although he can still understand what people are saying. Explain which area of his brain has been damaged. [2 marks]

I'm not sure that I have a brain...

It goes without saying that the brain is a pretty complicated organ. That doesn't mean that it has to be too stressful to learn about it for your exam. All you have to do is know what each of the main areas do, and understand that the brain processes information from receptors and sends impulses to effectors. Don't worry about exactly what's going on there.

The Autonomic Nervous System

These pages deal with the autonomic nervous system, which controls all the stuff in your body that happens without you thinking about it. If only revision happened that way...

The Autonomic Nervous System controls Unconscious Activities

The nervous system is divided into **two parts**:

1) **Conscious activity** is controlled by the **voluntary (somatic) nervous system.**
2) **Unconscious activities**, like the actions of the heart and the digestive system, are controlled by the **autonomic nervous system**, which sends impulses to the involuntary (smooth) muscle and glands. Unconscious, involuntary reactions to stimuli are 'reflex' reactions. They're stereotypic — there's always the same reaction to the stimulus.

The Autonomic Nervous System is divided into Two Parts

The autonomic nervous system is made up of the **sympathetic** and the **parasympathetic** nervous systems, which have **opposite effects** on organ activity.

These two systems can have **different effects** on the same kinds of muscles because the motor neurones involved secrete different kinds of neurotransmitter at the synapses (see page 26 for more on these).
The sympathetic nervous system uses **noradrenaline**, and the parasympathetic system uses **acetylcholine**.

The Sympathetic Nervous System causes the 'Fight or Flight' response

The **sympathetic nervous system** is the part of the **autonomic nervous system** that increases the overall physical activity of the body, a response called **'fight or flight'**. In a 'fight or flight' response blood is diverted from the gut to the **lungs**, **heart** and **voluntary muscle**:

1) It **increases** the heartbeat rate, to increase **oxygen supply** to the muscles.
2) It **increases** the ventilation rate, so that **more oxygen** can be absorbed in the lungs.
3) It **decreases** peristalsis in the gut so there's more blood available for the heart and lungs.

Other parts of the body also respond to enable **increased sensory awareness**. For example, the radial muscles of the iris contract, causing the pupils of the eye to dilate (see p. 22). The sympathetic nervous system triggers the same kinds of responses as **adrenaline**, the 'fight or flight' **hormone**.

The Parasympathetic Nervous System prepares the body for Rest

The **parasympathetic nervous system** is the part of the autonomic nervous system that decreases overall physical activity, so it's associated with rest:

1) It **decreases** heartbeat rate.
2) It **decreases** ventilation rate.
3) It **increases** peristalsis in the gut, so that food can be digested.

The Autonomic Nervous System

The **Sympathetic** and **Parasympathetic** systems have **Opposite Effects**

The two systems have a range of effects on the body. One example of the two systems working together with opposite effects is that **pupil diameter** can be changed to control the amount of light reaching the retina of the eye:

1) If there's **too much** light, the **parasympathetic** nervous system causes the circular muscles to contract, to **reduce** pupil diameter.
2) If light levels are **low**, the **sympathetic** nervous system contracts the radial muscles, **increasing** the diameter of the pupil.

The parasympathetic nervous system also controls tear production in the eye. If dust or something touches the cornea, lachrymal (tear) glands release tears to wash it out.

Comparison of the sympathetic and parasympathetic nervous systems

Feature	Sympathetic	Parasympathetic
transmitter substance at synapses	noradrenaline	acetylcholine
heart rate	speeds up	slows down
iris	dilates	constricts
movements of digestive tract	slows down	speeds up
sweating	stimulated	not stimulated

Bladder Control can be **Learned**

Some of the functions controlled by the autonomic nervous system can be **learned** over time, e.g. children learn bladder control:

1) The **sympathetic nervous system** causes a **sphincter** (a circular muscle) to contract at the bladder opening, letting the bladder **fill up**.
2) As the bladder becomes full, **stretch receptors** in the walls produce **nerve impulses** which go to the central nervous system (p.28).
3) The **parasympathetic nervous system** sends impulses to the sphincter, **relaxing** it.
4) The bladder **empties**.

Young children learn how to **recognise** the messages from the stretch receptors, and then control the sphincter **consciously** — this is potty training.

Harry wished that his parasympathetic nervous system would stop making such a mess on the floor.

Practice Questions

Q1 What's the difference between the autonomic nervous system and the voluntary (somatic) nervous system?
Q2 What is meant by the 'fight or flight' response?
Q3 Give three effects of stimulation of the body by the parasympathetic nervous system.

Exam Question

Q1 The size of the pupil of the eye decreases in bright light. This is an example of a reflex response and is brought about by the parasympathetic nervous system, which stimulates contraction of the circular muscles of the iris and effects relaxation of the radial muscles.

a) Explain why this kind of behaviour is described as a reflex. [2 marks]

b) Acetylcholine is the neurotransmitter released at the end of parasympathetic nerve fibres. Suggest how a single type of neurotransmitter can affect the different muscles in different ways. [2 marks]

No one will be sympathetic if you don't learn this...

*They're weird names, really. How can a nervous system be sympathetic? It doesn't offer you a tissue when you're upset. Basically, you just have to learn that the sympathetic system gets things ready for fight or flight, and the parasympathetic system calms things down. Try remembering it like this — **s**ympathetic for **s**tress, and **p**arasympathetic for **p**eace.*

Muscle Structure and Function

If the receptors in your eyes detect a hungry-looking lion coming towards you, your nervous system sends a message to your legs telling them to run away. This is where muscles come in handy. Read on...

Muscle** is a tissue made up of cells that are **Contractile

Contraction of a muscle gives a 'shortening force' that causes **movement** or sets up **tension**. The process needs lots of **energy** from ATP. There are three main types of muscle:

1) **Striated muscle** (also called **skeletal** or **voluntary muscle**) is attached to bone via **tendons** and is controlled by the motor neurones of the **voluntary nervous system**.

2) **Smooth muscle** (also called **unstriated** or **involuntary muscle**) is found in walls of tubular organs and is controlled by the **autonomic nervous system**.

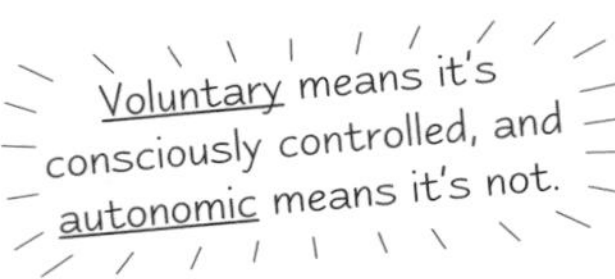

These two types of muscle only contract when stimulated by an **action potential** from the nervous system. They're **neurogenic**.

3) **Cardiac muscle** is found only in the **heart**. It's **myogenic**, which means it contracts **spontaneously**, without input from the nervous system. But the **autonomic nervous system** does control the **rate** of contraction.

Striated Muscle** is made up of **Muscle Fibres

Each **muscle fibre** contains lots of nuclei and is bound by a cell membrane called a **sarcolemma**. Contraction depends on protein **myofilaments**, which are arranged in bundles called **myofibrils**. The pattern of **thin** myofilaments (made of the protein **actin**) and **thick** myofilaments (made of the protein **myosin**) gives the **striations** (stripes) in the muscle.

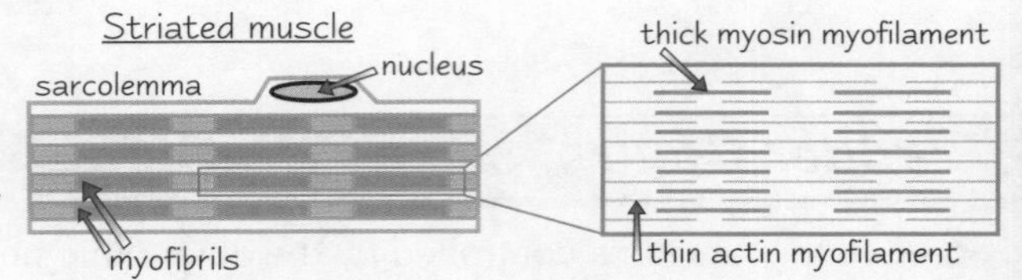

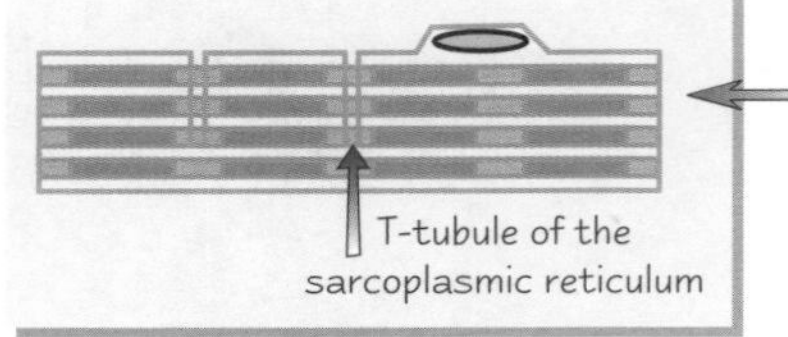

The cytoplasm of muscle fibres is called **sarcoplasm** and it's penetrated by **transverse (T) tubules** which make up a network called the **sarcoplasmic reticulum**. They let the sarcolemma transmit action potentials in towards the myofilaments. There are lots of **mitochondria** to provide ATP for contraction.

There are alternating dark and light bands in a muscle fibre. The dark ones are called **A bands** (think d**a**rk), and the light ones are called **I bands** (think l**i**ght). The middle bit of the A band, the **H zone**, is lighter than the rest of it, because there's no **overlap** between the myosin and actin myofilaments. A **Z line** connects the middle of the **actin** myofilaments, and an **M line** connects the middle of the **myosin** ones. The section of myofibril between two Z lines is called the **sarcomere**.

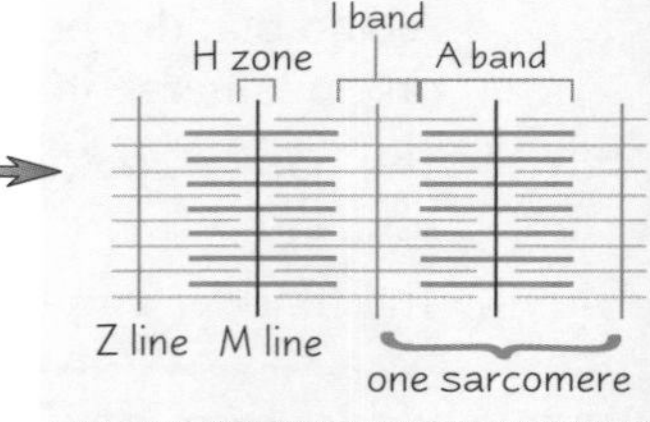

***Muscles Contract** when **Myofilaments** slide over one another*

Muscle contraction is explained by **Huxley's sliding filament hypothesis**.
It depends on bits of myosin called '**myosin heads**' binding to sites on actin filaments:

1) When muscle is stimulated it sets up an **action potential** in the **sarcolemma**, which spreads down the membranes of the **T-tubules**.
2) The **sarcoplasmic reticulum** membranes now become much more **permeable** and **calcium ions** diffuse out rapidly.
3) The Ca^{2+} quickly reaches the **actin filaments** and binds to a protein called **troponin**. This causes another protein called **tropomyosin** to change position and **unblock** the **binding sites** on the actin filaments.
4) ATP (from the myosin head) is hydrolysed and the energy released causes the myosin heads to alter their **angle** and attach to the binding sites forming **actomyosin cross bridges** between the two filaments.
5) The myosin head then **changes angle**, pulling the actin over the myosin towards the **centre** of the sarcomere.
6) ATP provides energy for the cross bridges to detach and then reattach, this time **further along** the actin filament. In effect the myosin heads '**walk**' along the actin filaments until they reach the end.

The process keeps repeating so that the whole muscle **contracts**.
This changes the width of the bands in the muscle (see next page).

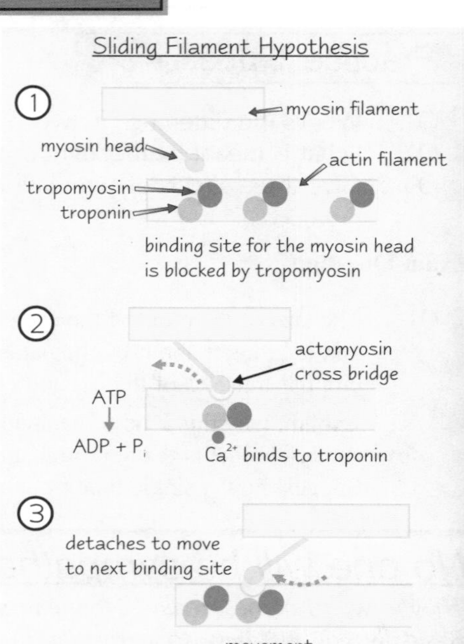

Muscle Structure and Function

Muscles **Relax** when **Excitation Stops**

When **excitation** of the sarcomere stops (i.e. the membrane has recovered its **resting potential**), calcium ions are **actively pumped** back out of the cytoplasm into the **T tubules**. This means that the troponin is released, and the tropomyosin moves back to **block** the myosin binding site again.

The cross bridges are broken and the myosin **detaches** from the actin filament. The sarcomere returns to its resting position and the width of the bands in the muscle change back to their relaxed appearance.

Relaxed muscle

Contracted muscle

actin and myosin filaments overlap more

Muscles have many Features like those of **Neurones**

1) When a muscle fibre is stimulated by an action potential, it only contracts if the stimulus is above a certain **threshold level**.
2) Muscle fibre membranes have an **all-or-nothing** nature, and so increasing the strength of a stimulus above the threshold level doesn't give a bigger force of contraction in the muscle fibre. But an increase in stimulus strength does increase the **number of muscle fibres** that contract.
3) Bigger forces of contraction happen in muscles with bigger **cross-sectional areas**.
4) After the response, the muscle goes through an **absolute refractory period** (when no contraction can happen), followed by a **relative refractory period** (when it takes a **bigger stimulus** to get a response). The refractory period is the time it takes for the muscle to go back to its **resting potential** and be capable of further stimulation.
5) As muscles can only produce a **shortening force**, at least **two** sets of muscles have to be used to move a bone into position and back again. Pairs of muscles acting in this way are called **antagonistic pairs**.

Muscles work in **Antagonistic Pairs**

You need **muscles** to move your bones.
Muscles are attached to bones by **inelastic tendons**.
The tendons can't stretch, so when a **muscle contracts** it **shortens** and **pulls** the bone.
A different muscle returns the bone to its original position.
Muscles are usually found in **antagonistic pairs** — in the arm, when the **biceps contracts**, it pulls the lower arm upwards. When the **triceps contracts** the lower arm goes back to its original position. This is a **lever action**.

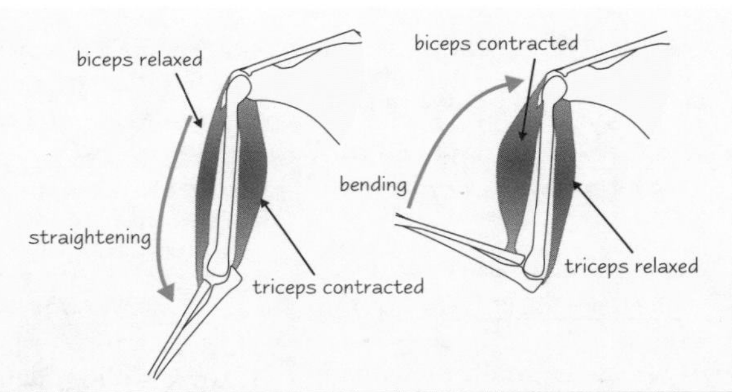

Practice Questions

Q1 Name the three main types of muscle, and say which part of the nervous system controls each one.

Q2 Which type(s) of myofilament (thick or thin) are present in the following bands on a muscle fibre:
a) A band b) I band c) H zone?

Q3 What is the refractory period of a muscle?

Exam Questions

Q1 Describe the sequence of events that lead to the contraction of muscle following the arrival of an action potential along the T tubule. [10 marks]

Q2 Predators have many muscle fibres with particularly well developed sarcoplasmic reticulum. Suggest the advantage of this. [5 marks]

Muscles — pretty, but very boring...

Similar to models in that respect. Not that I know many models, so I'm probably being very unfair. Probably many models have degrees in psychology or philosophy, and enjoy sky-diving and oil-painting in their spare time. Male models are often quite muscular, which leads us neatly back to the original subject. You need to learn everything on this page about muscles.

Meiosis

You might remember meiosis from AS — it's the one needed for sexual reproduction (no sniggering at the back, please).

Meiosis is a Special type of Cell Division

Meiosis **halves** the chromosome number. It's used for sexual reproduction in plants and animals.

In animals, meiosis produces the **gametes** (sperm and egg cells) and takes place in the **testes** and **ovaries**. When the **gametes** fuse at **fertilisation** they combine their chromosomes, so the chromosome number is **restored**. These two processes make sure that **chromosome numbers stay constant** overall from generation to generation.

Meiosis creates Haploid Cells

Meiosis has **two divisions**, and each one is made up of separate stages. Don't worry about learning the names or exact details of each stage, but do learn about what happens to the chromosomes. The **first division** (parts 1-5 in the diagram) **halves** the chromosome number, and the **second division** (parts 6-11) **separates** the pairs of chromatids that make up each chromosome, like in **mitosis**.

1. INTERPHASE

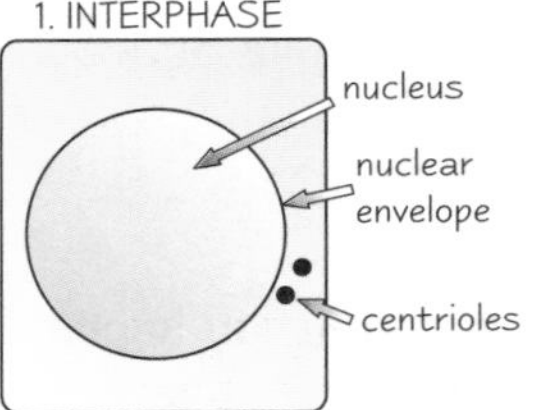

The cell is diploid (it has two sets of chromosomes.) It replicates its DNA, ready to divide.

2. PROPHASE I

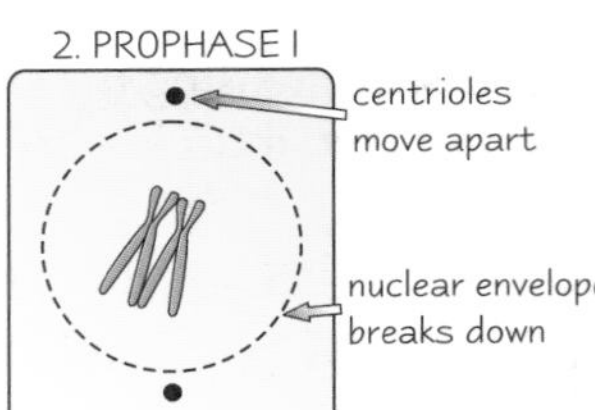

Chromatin coils up and becomes visible as chromosomes, each chromosome consisting of 2 chromatids. Homologous chromosomes pair up — the pairs are known as bivalents.

3. METAPHASE I

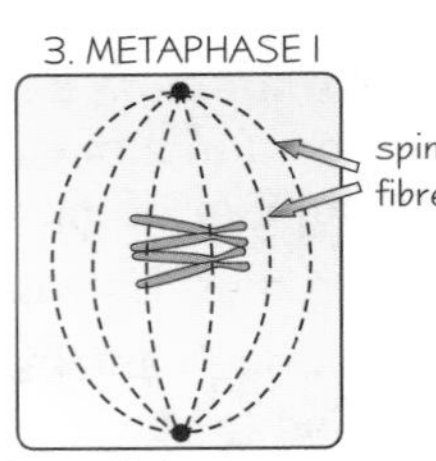

Pairs of chromosomes go to the centre of the cell.

4. ANAPHASE I

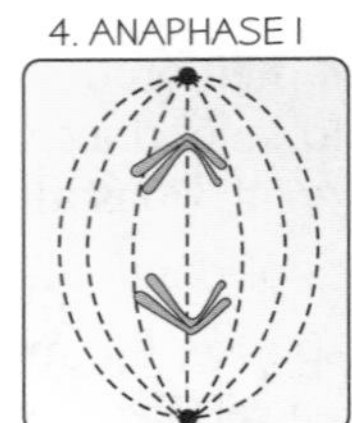

Members of each homologous pair of chromosomes separate and are pulled apart along the spindle fibres.

5. TELOPHASE I

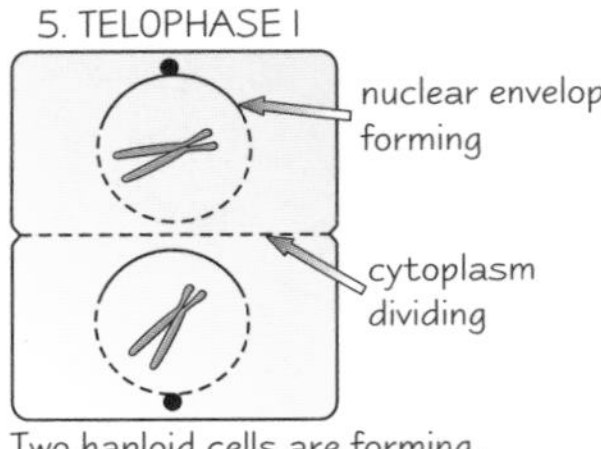

Two haploid cells are forming.

6.

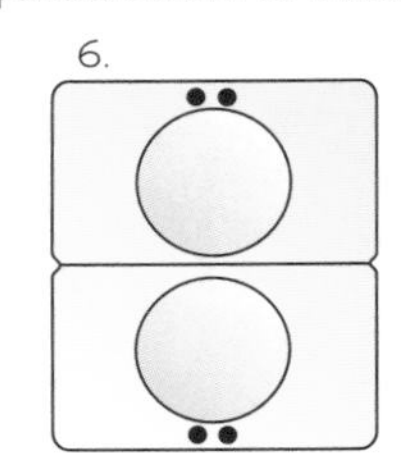

This is the beginning of meiosis II. The cells prepare to divide again.

7. PROPHASE II

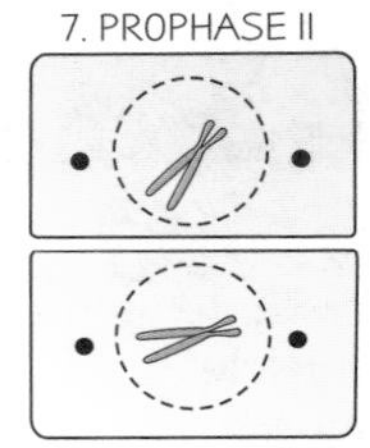

Chromosomes coil up once more. Each chromosome consists of 2 sister chromatids.

8. METAPHASE II

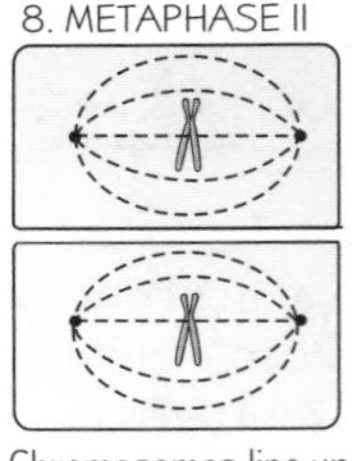

Chromosomes line up in centre of cells.

9. ANAPHASE II

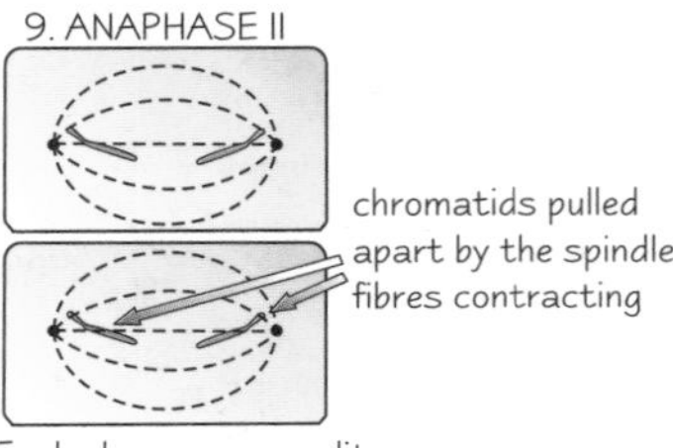

Each chromosome splits into its chromatids.

10. TELOPHASE II

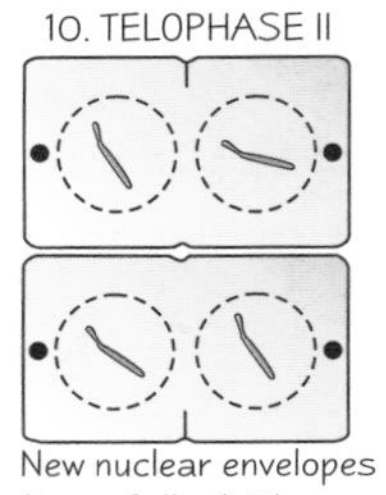

New nuclear envelopes form. Cells divide.

11. FOUR HAPLOID CELLS (e.g. sperm cells)

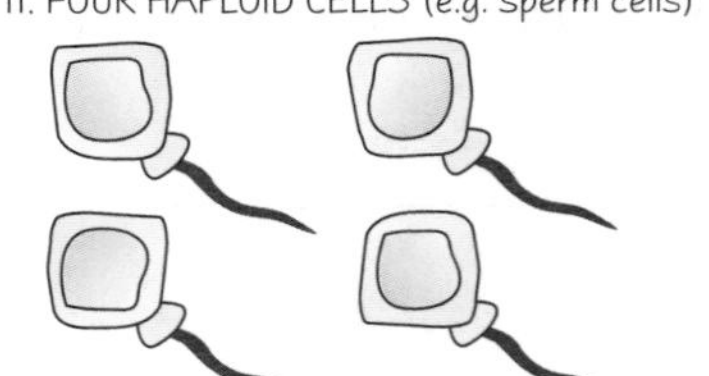

Haploid cells have only one set of chromosomes.

The diagram only shows what happens to one pair of chromosomes. In human cells, there are 23 pairs of chromosomes in total, all doing the same thing.

Cells divide Twice in Meiosis

First division (Meiosis I)

The chromosome pairs come together. The two chromosomes of a pair are called **homologous** chromosomes (see the next page). When they pair up, the pair is called a **bivalent**. Then, these homologous chromosomes move to opposite ends of the cell, and the cell divides. Now, there are **two haploid cells** instead of one diploid cell.

Second division (Meiosis II)

This is **similar** to **mitosis** (check in your AS notes if you can't remember). Each new **haploid cell** divides, and each chromosome splits into its **chromatids**.

Meiosis

Crossing Over happens between Chromatids

During **prophase I**, the homologous chromosomes **exchange** pieces of their chromatids. This is called **crossing over.**

Crossing over happens randomly between the homologous chromosomes at any place along them.

The place where crossing over occurs is called a **chiasma** (plural: **chiasmata**). Crossing over helps to mix up alleles in new combinations and creates **variation**.

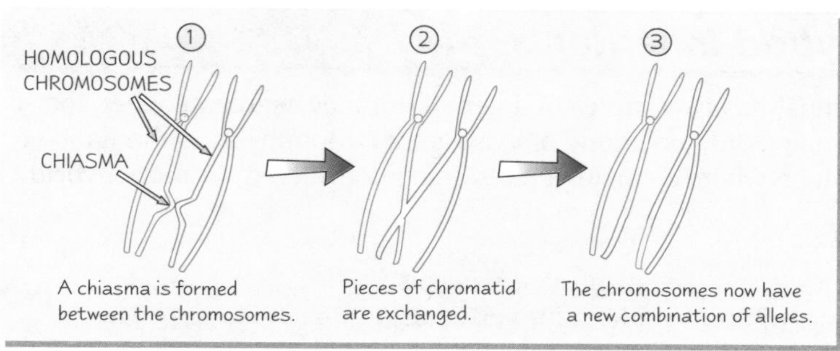

A chiasma is formed between the chromosomes.

Pieces of chromatid are exchanged.

The chromosomes now have a new combination of alleles.

You need to Learn some Key Terms

There's a lot of **fancy words** used about genetics and inheritance. Make sure you know these **important ones**:

chromosome	A strand of genetic material (DNA) found in the nucleus of a cell. Each chromosome consists of one molecule of DNA and histone proteins.
chromatid	One of the two identical strands of genetic material that make up a chromosome during cell division.
homologous	Homologous chromosomes are a pair of equivalent chromosomes with the same structure and arrangement of genes — usually one is inherited from the mother and one from the father.
bivalent	A pair of homologous chromosomes.
haploid	Where a cell contains only one complete set of chromosomes, e.g. sperm and egg cells.
diploid	Where a cell contains two complete sets of chromosomes — in pairs of homologous chromosomes.
gene	A section of DNA on a chromosome which controls a characteristic in an organism. It carries the genetic code to make one or more polypeptide or protein, by coding for RNA.
locus	The position on a chromosome where a particular gene is located.
allele	An alternative form of a gene. E.g. in pea plants, the gene for height has two forms — one allele for tall plants and one allele for short plants.
genotype	The alleles a particular individual has.
phenotype	An individual's characteristics (e.g. eye colour, height), which result from the interaction of genotype and environment.
homozygous	An individual with two copies of the same allele for a particular gene.
heterozygous	An individual with two different alleles for a particular gene.
dominant	The condition in which the effect of only one allele is apparent in the phenotype, even in the presence of an alternative allele.
codominance	The phenomenon in a heterozygote in which the effects of both alleles are apparent in the phenotype.
recessive	The condition in which the effect of an allele is apparent in the phenotype of a diploid organism only in the presence of another identical allele.
linked	Genes located on the same chromosome that are often inherited together.

Practice Questions

Q1 Place these events in meiosis in the correct order: A. chromatids separate; B. homologous chromosomes pair up; C. two haploid cells are produced; D. homologous chromosomes separate; E. four haploid cells are produced.

Q2 What's the difference between genotype and phenotype?

Exam Questions

Q1 In which organs of the human body does meiosis occur? [2 marks]

Q2 Explain the difference between: a) a gene and an allele; b) haploid and diploid. [4 marks]

Q3 Explain the importance of meiosis in the life-cycles of sexually reproducing organisms. [3 marks]

How do you tell the sex of a chromosome? Take down its genes...

Remember that genes are carried on chromosomes, so whatever the chromosomes do (like separating and re-combining), the genes will do too. It's a huge diagram, but just break it down into meiosis I and II, and learn what happens in general. Don't worry about learning the names of stages though — they're just mentioned here in case you're interested.

Inheritance

Brace yourself for two pages of genetic diagrams. You need to get comfortable with these, because in the exam you'll not only have to interpret them, you might have to draw some of your own. It's probably a smart idea to learn some of the most common patterns and ratios, then you'll be able to apply them to new examples in the exam.

Monohybrid Inheritance Involves One Characteristic

Each individual has **two copies** of a gene. But they **segregate** when the sex cells are formed in meiosis, so each **gamete** contains only **one copy** of **every** gene. Monohybrid inheritance is the **simplest** form of inheritance — it's just inheritance where a single gene is being considered. A **monohybrid cross** is a genetic cross for only one gene:

Example

In fruit flies, the allele for **normal wings** is **dominant** (N), and the allele for **vestigial** (short) wings is **recessive** (n).

A normal-winged fruit fly is crossed with a fruit fly that has vestigial wings. **All** the offspring are normal-winged. These flies then **interbreed**, and the next generation shows a **3:1 ratio** of normal wings to vestigial wings, i.e. a 75% chance of normal wings and a 25% chance of vestigial wings.

The first set of offspring from an experiment like this, where the two parents are true-breeding (homozygous), is called the F_1 generation. If you then breed these offspring together, you produce the F_2 generation.

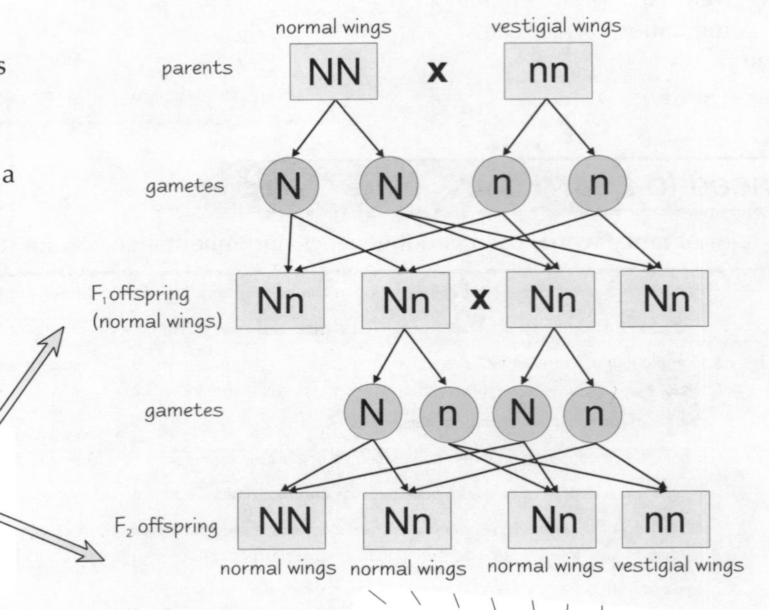

Genetic diagrams like this show all the possible combinations of gametes from the parents. Remember to use a capital letter for a dominant allele, and a small letter for a recessive allele.

A Test Cross helps you find out an Individual's Genotype

Sometimes, you might cross a normal-winged fly with a vestigial-winged fly and get a **1:1 ratio** of normal wings to vestigial wings in the offspring, instead of all normal. This happens if the normal-winged fly is **heterozygous** — it has **one allele** for **normal** wings, and **one allele** for **vestigial** wings. Because the allele for vestigial wings is recessive, it doesn't show up in the phenotype of heterozygous flies — vestigial wings is a recessive condition.

Compare this with the first diagram on this page. In each case, the normal fly **looks** the same (they have the same **phenotype**). The only way of telling its genotype is by a **breeding experiment** where you mate it with a recessive individual — in this case that's a fly with **two alleles** for **vestigial** wings (remember this because it'll crop up in the exam). This is called a **test cross**.

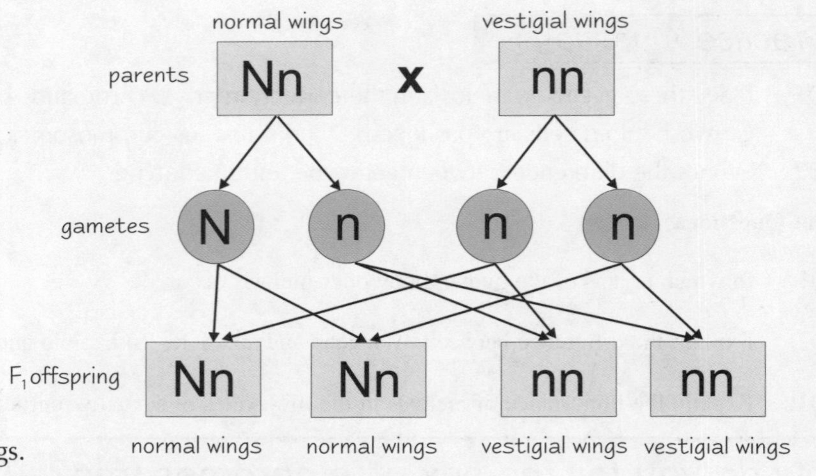

- a **homozygous** normal-winged fly produces **all** offspring with the **dominant** characteristic when it's crossed with a fly with vestigial wings.
- a **heterozygous** normal-winged fly produces about **half** the offspring with the **recessive** characteristic when it's crossed with a fly with vestigial wings.

Inheritance

Alleles can be Codominant

Occasionally, alleles show **codominance**.
One example in humans is the allele for **sickle-cell anaemia**:

- Normal people have two alleles for normal haemoglobin (**H^NH^N**).
- People with **sickle-cell anaemia** have two alleles for the disease (**H^SH^S**). They have abnormal haemoglobin, which makes their red blood cells sickle-shaped and unable to carry oxygen properly. Sufferers usually die quite young.
- Heterozygous people (H^NH^S) have an in-between phenotype, called the **sickle-cell trait**. Some of their haemoglobin is normal and some is abnormal, but the red blood cells are normal-shaped. The two alleles are **codominant**, because they're **both** expressed in the **phenotype**.
- The sickle-cell allele is a result of a **mutation**.

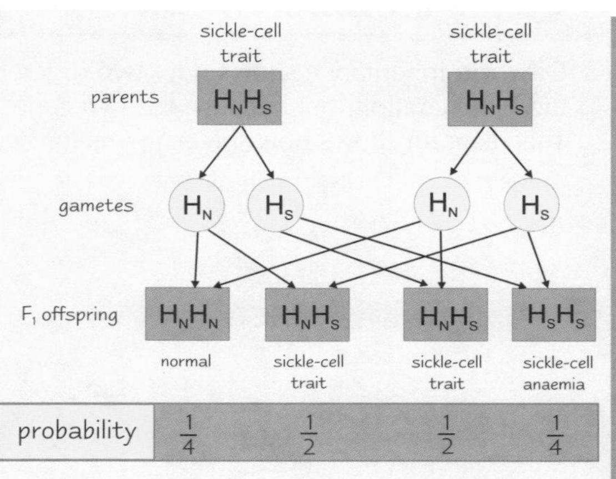

Some Genes have Multiple Alleles

Inheritance is more complicated when there are more than two alleles of the same gene — **multiple alleles**.
E.g. in the **ABO blood group system** there are **three alleles** for blood type:

> I^O is the **recessive** allele for blood group **O**. I^A is the allele for blood group **A**. I^B is the allele for blood group **B**.

Alleles I^A and I^B are **codominant** — people with copies of **both** these alleles will have a **phenotype** that expresses **both** alleles, i.e. blood group **AB**. In the diagram below, a couple who are both **heterozygous**, one for blood group A and one for blood group B have children, those children could have one of **four** different blood groups — A, B, O or AB.

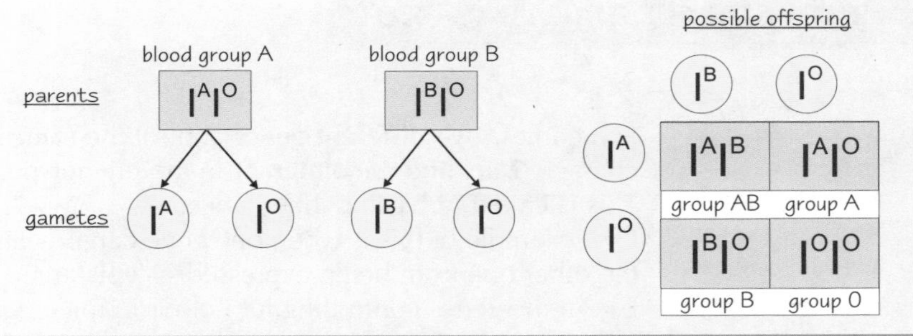

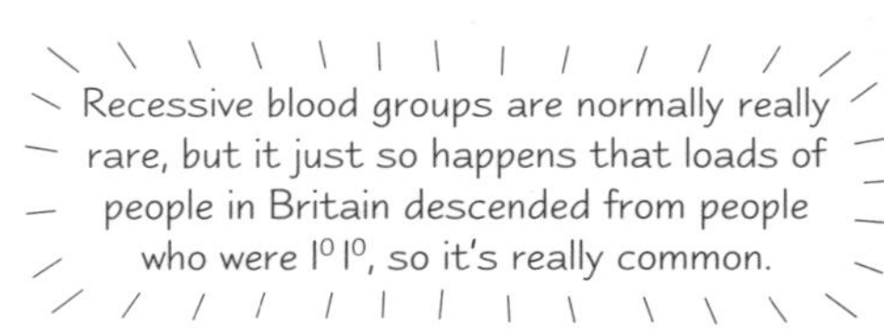

Practice Questions

Q1 What genetic ratios do you expect from each of these crosses?
a) Aa × Aa b) AA × Aa c) Aa × aa

Q2 Do the genetics of the ABO blood group system show multiple alleles, codominance, or both?

Q3 What is sickle-cell anaemia? What type of inheritance pattern do sickle-cell alleles show?

Exam Questions

Q1 List the six possible genotypes for the human ABO blood groups. [3 marks]

Q2 In pea plants, the allele for purple flowers is dominant over the allele for white flowers.
How would you find out if a purple-flowered plant is homozygous or heterozygous? [3 marks]

It's hard to do test crosses on humans...

If you're wondering whether you're heterozygous for a particular trait, it's probably not an option to breed with a recessive person, and then have lots of babies and see what they look like, unless you take your science homework very seriously.

Inheritance

There's so much to say about inheritance that we've generously stuck in another two pages for you to enjoy.

Genes on **Different Chromosomes** Segregate **Independently**

Dihybrid inheritance shows how **two** different genes are inherited. Each gene gives a 3:1 ratio in the F_2 generation, but because the two genes do this **independently**, it makes a **9:3:3:1 ratio** overall. This diagram shows how this happens for two traits in the fruit fly.

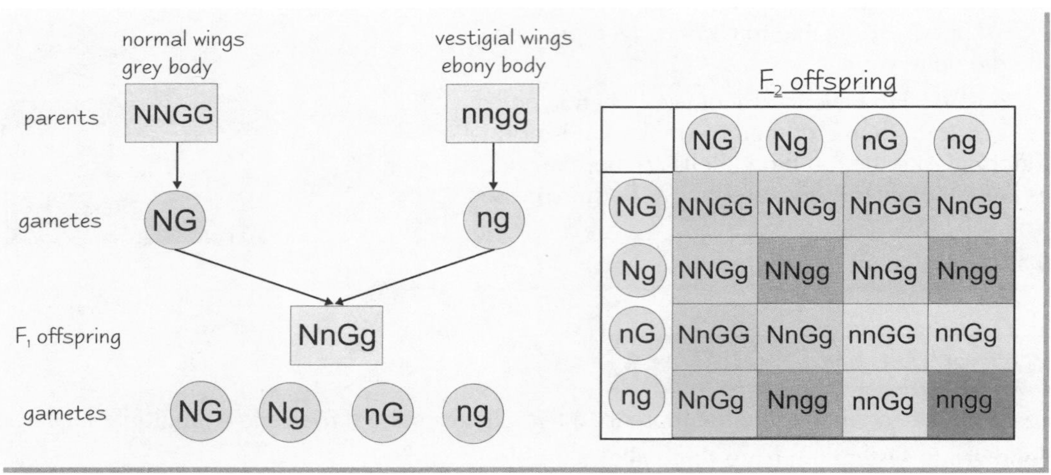

	NG	Ng	nG	ng
NG	NNGG	NNGg	NnGG	NnGg
Ng	NNGg	NNgg	NnGg	Nngg
nG	NnGG	NnGg	nnGG	nnGg
ng	NnGg	Nngg	nnGg	nngg

Crossing an F_1 fly with a double recessive fly (vestigial wings and ebony body) gives a **1:1:1:1 ratio**. Check your understanding by working this out yourself.

An **Epistatic** gene affects the **Expression** of another Gene

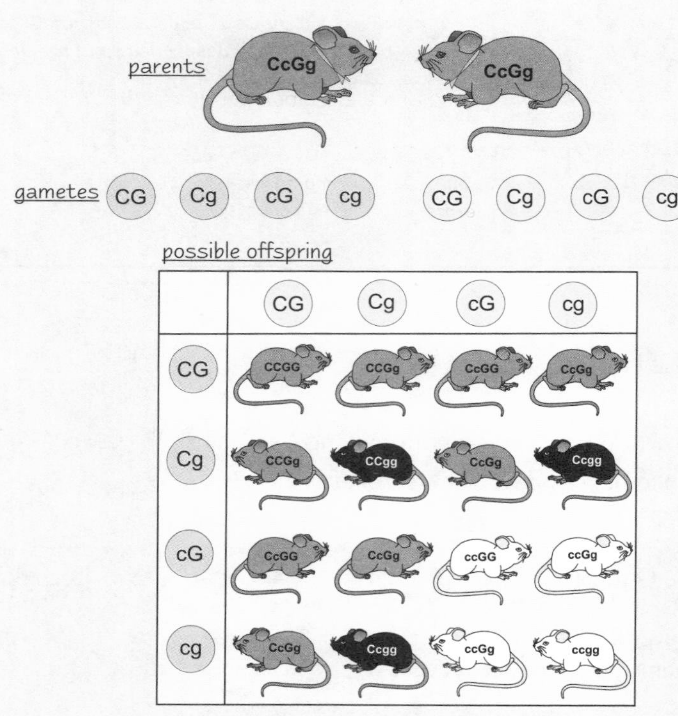

Sometimes, **two different genes** control the **same** characteristic, and they **interact** in the phenotype. This is called **polygenic** inheritance.
One example of this is when one gene can prevent the other one from being expressed — **epistasis**. E.g. in the genes controlling fur colour in mice:

1) **One gene** (C/c) decides whether the fur is **coloured** (C) or **albino** (c).
2) This gene is **epistatic** over a **second** gene (G/g) which makes the colour (if any) **grey** or **black**. The **expression** of the second gene is affected by the first gene.
3) If a mouse is **recessive** for the first gene (cc) then the mouse will be **albino** and the second gene doesn't have any effect on the phenotype.

Epistasis can produce some **weird ratios** when you start crossing heterozygous individuals together — in the case of the mice, it produces the ratio **grey: 9, black: 3, albino: 4.**

Inheritance

In Mammals Sex is Determined by the X and Y Chromosomes

The genetic information for your **gender** is carried on two **specific** chromosomes:

1) In mammals, **females** have **two X** chromosomes, and **males** have **one X** and **one Y**. The probability of having male or female offspring is **50%**.

2) The Y chromosome is **smaller** than the X chromosome and carries **fewer genes**. So most genes carried on the sex chromosomes are only carried on the X chromosome. These genes are sex-linked. Males only have **one copy** of the genes on the X chromosome. This makes them more likely than females to show **recessive phenotypes**.

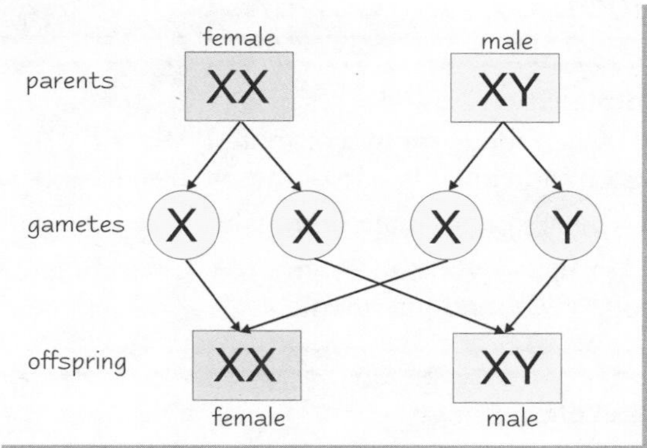

3) Genetic disorders inherited this way include **red-green colour-blindness** and **haemophilia**. The pattern of inheritance can show that the characteristic is **sex-linked**. In the example below, females would need **two copies** of the recessive allele to be colour-blind, while males only need one copy. This means colour-blindness is **much rarer** in **women** than **men**.

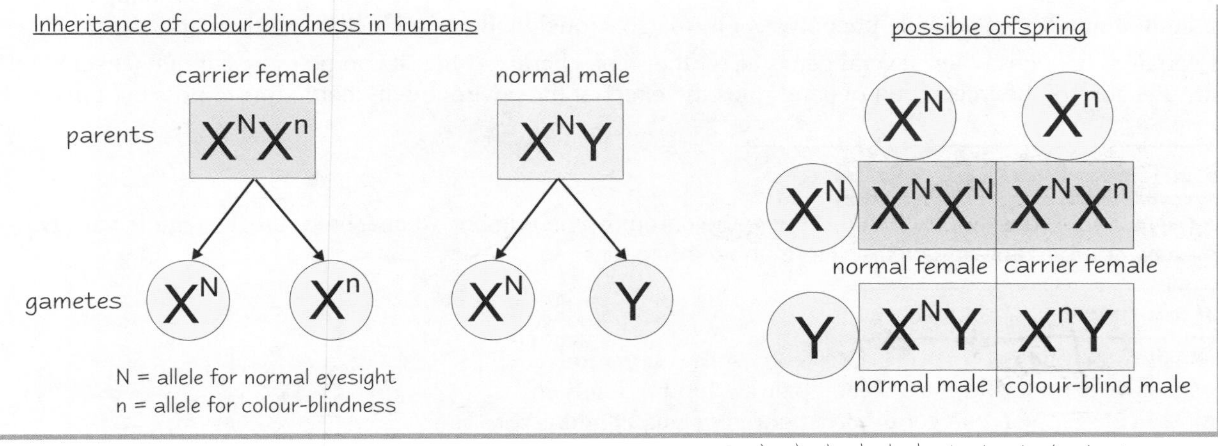

A carrier is a person carrying an allele which is not expressed in the phenotype, but which can be passed on.

Practice Questions

Q1 What does epistasis mean?

Q2 Which chromosomes determine the gender of mammalian offspring?

Q3 Why is red-green colour-blindness much more common in males than in females?

Exam Questions

Q1 Draw a genetic diagram to show the expected results of a cross between a normal-winged grey-bodied fruit fly (genotype NnGg) and a normal-winged ebony-bodied fruit fly (Nngg). [4 marks]

Q2 The recessive allele for haemophilia is carried on the X chromosome. Explain why you would expect haemophilia to be more common in males than females. [5 marks]

Pedigree charts aren't just for dogs... they're for royals too...

You can use pedigree charts to track the inheritance of sex-linked conditions like haemophilia. The classic example is the inheritance of haemophilia from Queen Victoria, who carried the allele and spread it through the royal families of Europe.

Variation

Ever wondered why no two people are exactly alike? No, well nor have I, actually, but it's time to start thinking about it. This variation is partly genetic and partly due to differences in the environment.

Variation can be Continuous or Discontinuous

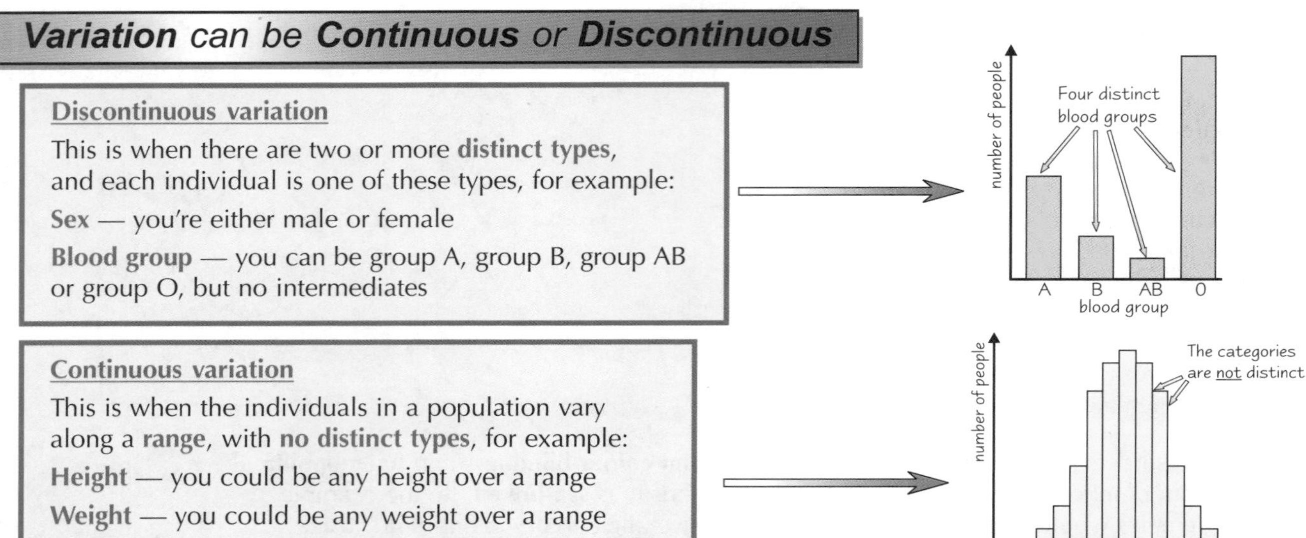

Discontinuous variation

This is when there are two or more **distinct types**, and each individual is one of these types, for example:

Sex — you're either male or female

Blood group — you can be group A, group B, group AB or group O, but no intermediates

Continuous variation

This is when the individuals in a population vary along a **range**, with **no distinct types**, for example:

Height — you could be any height over a range

Weight — you could be any weight over a range

Skin colour — any shade from very dark to very pale

Discontinuous variation has clear-cut categories because it depends on only one or a few genes (it is **monogenic**). So, there's a **limited number** of possible phenotypes. It isn't so strongly influenced by the environment.

Continuous variation happens when **several** genes affect the same characteristic. It can be more strongly affected by the **environment**. Because of the interaction of genes plus the effect of the environment, there's lots of possible **phenotypes**.

Meiosis affects Genetic Variation

Meiosis (see pages 34-35) does more than just halve the chromosome number. It also helps create **genetic variety**, by producing new combinations of alleles. Here's how it happens:

Independent assortment of chromosomes

During **meiosis I**, the pairs of homologous chromosomes **separate** (at anaphase). The chromosomes from each pair end up **randomly** in one of the new cells, so you can get **different chromosome combinations**. In **meiosis II**, there's also random assortment of **chromatids**.

One pair of chromosomes would give **2** different types of haploid cell.

- Two pairs would give 2^2 possible haploid cells = **4 possibilities**.
- 23 pairs, like in humans, give 2^{23} possible haploid cells = over **8 million possibilities**. (Your parents would have to have millions of children before they stood any chance of having two genetically the same — unless they have twins.)

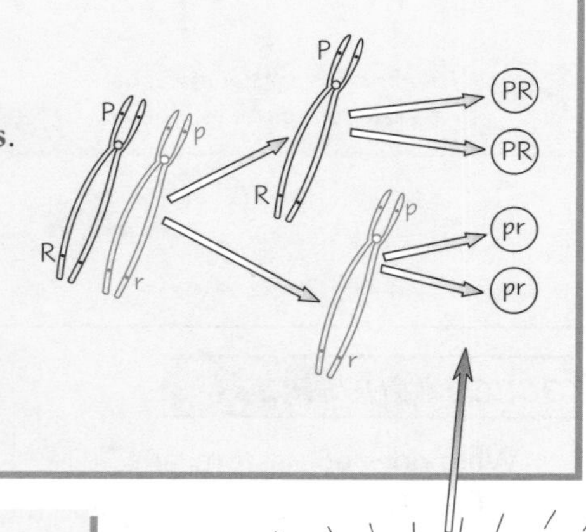

Crossing Over

Chromosomes often swap parts of their chromatids during **prophase I** (see page 35). This creates **new combinations** of alleles on those chromosomes, separating alleles that are normally inherited together (**linked**).

PR
Pr
pR
pr

Meiosis also produces variation because it lets **fertilisation** take place. The **random fusion** of gametes from two individuals at fertilisation creates unique **combinations** of alleles.

Mutation

Mistakes sometimes happen during cell division, producing a **completely new** characteristic (see page 42).

Variation

Environment affects **Phenotype**

A lot of variation in characteristics (phenotype) is due to differences in genotype, but **environment** also has an effect:

1) The **Himalayan rabbit** is mainly white, but some parts of its fur (at the ears, feet and tail) are black. The growth of the black fur is caused by environmental temperature — these parts of the body are cooler, and the black colour only develops when the skin temperature is below about 25 °C.
2) People are on average much **taller** today than they were 200 years ago (if you're a strapping six-footer, you'll probably bump your head on the ceilings of an old house). This is thought to be because our diet is much better.
3) **Plant growth** is strongly affected by the environment — plants show better, healthier growth when there are more **nitrates** and other minerals available in the soil.

Genes and **Environment** interact in the **Phenotype**

Pea plants provide a clear example of the **interaction** between genes and environment that produces a **phenotype**.

Pea plants come in tall and dwarf forms. This characteristic (tall or dwarf) is passed on from one generation to the next, so we can tell that it is **genetic**.

However, the tall plants vary in height, and so do the dwarf plants, so **environment** is involved too.

> Tall or dwarf is **discontinuous** variation. Height variation among the plants of each type is **continuous** variation.

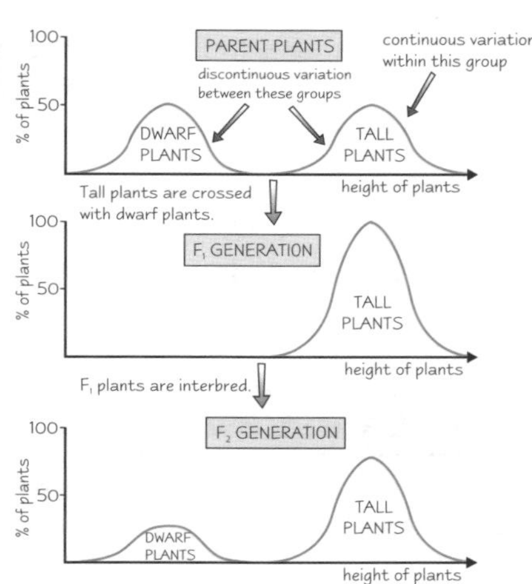

Twins can show the effects of **Genes** and **Environment**

Studies of **twins** have been used to find out if a human characteristic is mainly influenced by **genes** or by **environment**.

Monozygotic twins ("identical twins") have **identical genes** and **alleles**, because they both developed from the **same fertilised egg**. This means that if there are any differences in their characteristics, they must be due to the **environment**. Occasionally, monozygotic twins are raised **separately**, and comparing differences between them (compared with twins raised together) could show how important these environmental influences are.

Practice Questions

Q1 Give two examples of characteristics that show continuous variation, and two that show discontinuous variation.

Q2 State three ways in which meiosis helps to create variety within a species.

Q3 Describe the relationship between the effect of genetic and environmental variation on phenotype.

Exam Questions

Q1 If the body cells of an organism contain three pairs of chromosomes, how many different chromosome combinations can be produced in the gametes of this organism as a result of independent assortment? [2 marks]

Q2 Compare characteristics showing continuous and discontinuous variation with reference to:
a) the extent to which they are affected by the environment. [1 mark]
b) the number of genes that control them. [1 mark]

Variety is the spice of... meiosis...

By now you should have a pretty good idea of how meiosis creates variety in species. It's amazing to think of how many things influence the way that we look and behave. It's the reason we're all so lovely and unique... my parents often said they were glad they'd never have another child quite like me — I can't imagine why.

Natural Selection and Evolution

Darwin's book on his theory of evolution 'On The Origin of Species' is probably the most important biology book that's ever been published. Apart from this one of course.

Darwin wrote his Theory of Evolution in 1859

In **1831**, Darwin was invited to join the ship **HMS Beagle** on a map-making trip around the world. Darwin was a keen **naturalist** and on the voyage he collected lots of **data** and many **samples** of **plants** and **animals**. When he returned, he spent twenty years studying his data and eventually he came up with the **theory of evolution** which was published in **1859** in a book called '**On The Origin of Species by means of Natural Selection**'.
Darwin's theory was based on five main ideas:

1) **More individuals** are **produced** than can **survive**.
2) So, there is a **struggle** for **existence**.
3) Individuals within a species show **variation**.
4) So, those with **advantageous features** have a greater chance of **survival**.
5) Those individuals who **survive** produce **similar offspring**.

I'm not sure I'd like to try and sail in a beagle

Darwin used the term '**natural selection**' to describe the way that individuals with variations that help them survive in their habitat have advantages which make it more likely that they'll be able to pass on their genes. Darwin believed that natural selection **caused evolution**.

The Latest evolutionary theories still include Natural Selection

Obviously, there has been a lot more scientific research into natural selection since Darwin published his book.
The latest developments in genetics have been incorporated into Darwin's theory to update it:

1) There are changes in the **genetic composition** (gene frequencies) of a population from one generation to the next.
2) These changes are brought about by **mutation**, **genetic drift**, **gene flow** and **natural selection**:

Mutation

The **mutation** of genes can produce different **allele** and **phenotype frequencies**.

Genetic Drift

This is the **alteration** of **gene frequencies** through **chance**. For example, if two **heterozygous** individuals breed, their offspring might not be produced in an exact **Mendelian** ratio — the gene frequencies in populations change over time.

Gene Flow

This happens when new genes enter or leave a population by **migration**.

Natural Selection

1) As **conditions** change, or organisms **move** into a new environment, the organisms that are better **adapted** to the new conditions because of the **alleles** they carry will **survive**.
2) **Variation** between **isolated populations** increases as **gene pools diverge**.
3) Changes occur in **allele**, **genotype** and **phenotype** frequencies.
4) Eventually a **new species** will evolve.

Natural Selection and Evolution

Limiting Factors affect Survival and Reproduction Rates

All organisms tend to **overproduce** — this inevitably brings about **intraspecific competition** for resources. **Limiting factors** like **parasites** put **selection pressures** on the organisms. This is when natural selection comes into play — the individuals that are best adapted to the conditions because of the genes that they carry are more likely to survive and reproduce. This process changes the **allele frequencies** in the population. For example, there would be a **greater number** of **individuals** with **resistance** to certain **parasites** in the population.

Peppered Moths are a classic **Natural Selection** example

The **peppered moth** (*Biston betularia*) has **two phenotypes** — a **peppered** form and a **melanic** (black) form. The moth lives on the bark of trees. In areas with lots of **industrial pollution**, soot was deposited on the tree trunks, darkening them.

In these industrial areas, the **melanic** form became much more common — that's because the peppered form was much more obvious to predators, so it was far less likely to **survive** and **reproduce**. In areas where there was **little** pollution, the **peppered** form was more common.

Peppered and melanic moths on tree bark in unpolluted area

Peppered and melanic moths on tree bark in polluted area

Practice Questions

Q1 Write a paragraph explaining natural selection in your own words.

Q2 How has modern biology updated Darwin's theory of evolution?

Q3 What does genetic drift mean?

Exam Question

Q1 In 1976-1977, a severe drought struck the Galápagos islands. No rain fell for over a year. During the drought a number of plant species died out. Some others did not produce seeds, causing a food shortage for the seed-eating ground finch *Geospiza fortis.*

The seeds of one plant species that survived the drought were stored in large, tough fruits. Only *Geospiza fortis* individuals with a beak depth greater than 10.5 mm were able to feed from this plant.

A biologist conducted a survey of the finches and recorded the mean beak depths and lengths of birds that survived and birds that died during the drought. The results of the survey are recorded in this table:

	Beak Length	Beak Depth
Surviving birds	11.07	9.96
Dead birds	10.68	9.42

As the population of *G. fortis* recovered after the drought, the mean beak depth of the population was greater than before (an increase of 4–5%). Explain this change with reference to evolutionary change. [6 marks]

Lack of chocolate is a revision-limiting factor...

Natural selection is the most important thing on this page so make sure that you read that bit thoroughly. You need to be able to describe how natural selection happens and what impact it has on future generations of a species. Also, check that you know how genetics has been incorporated into modern thinking on evolution.

Speciation

Speciation is all about how new species appear — quite simple really, and surprisingly interesting. Read on...

Speciation is the Development of a New Species

A **species** is defined as a group of organisms that can **reproduce** and produce **fertile young**. Every true species identified so far has been named using the **binomial system** (see page 47).

Sometimes two individuals from different species can breed and produce offspring. These **hybrid** offspring aren't a new species because they're **infertile**. For example, **lions** and **tigers** have bred together in zoos to produce **tigons** and **ligers** but they aren't new species because they **can't** produce offspring.

Speciation (development of a new species) happens when **populations** of the **same species** become **isolated**. Local populations of a species are called **demes** (be careful with this — demes aren't always isolated).

Geographical Isolation causes Allopatric Speciation

1) Geographical isolation happens when a **physical** barrier **divides** a population of a species.
2) **Floods, volcanic eruptions** and **earthquakes** can all lead to barriers that cause some individuals to become **isolated** from the main population.
3) **Conditions** on either side of the barrier will be slightly **different**. For example, there might be a different **climate** on either side of the barrier.
4) Environmental conditions like this put **pressure** on the organisms, forcing them to **adapt** — the **natural selection processes** differ in each isolated group.
5) **Mutations** will take place **independently** in each population and, over a **long** period of time, the gene pools will **diverge** and the **allele frequencies** will **change**.
6) Eventually, individuals from different populations will have changed so much that they won't be able to breed with one another to produce **fertile** offspring — they'll have become **two separate species.**

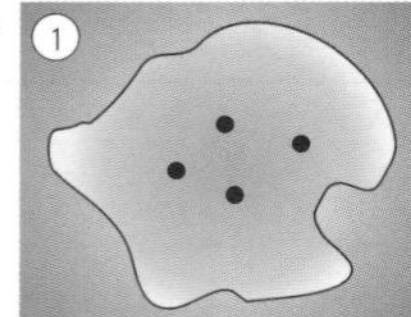

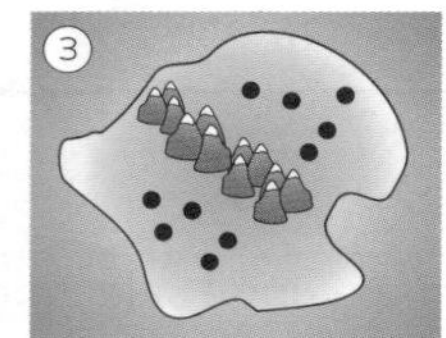

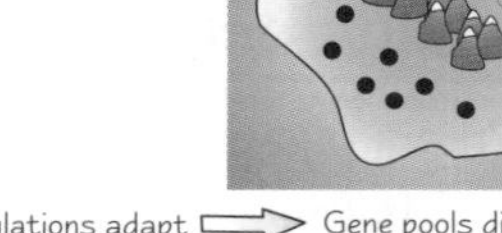

Population of individuals ⇒ Populations separate. ⇒ Physical barriers stop interbreeding between populations. ⇒ Populations adapt to new environments. ⇒ Gene pools diverge, leading to development of new species.

Isolation doesn't have to be Physical

Reproductive isolation happens when something **prevents** some members of a population breeding with each other. There are **several causes** of reproductive isolation:

1) **Seasonal isolation** — where mutation or genetic drift means that some individuals of the same species have different **flowering** or **mating** seasons, or become **sexually active** at different times of the year.
2) **Mechanical isolation** — where mutations cause changes in genitalia which prevent successful mating.
3) **Behavioural isolation** — where a group of individuals develop **courtship rituals** which are **not attractive** to the main population of a species.
4) **Gametic isolation** — where mutations mean that male and female **gametes** from different populations of the same species are **not** able to create new individuals — so the individuals can mate, but fertilisation fails or the foetus is aborted.

If two populations have become so different that they can't breed then a **new species** will have been created — this is called **sympatric speciation.**

Plant Speciation can occur through Polyploidy

Sometimes chromosomes don't separate during meiosis, and some gametes end up being **diploid** rather than **haploid**. If these gametes fuse with other gametes you end up with individuals that have one or more **extra sets of chromosomes** — that's **polyploidy**. Sometimes, the chromosome set doubles **after fertilisation** — the chromosomes replicate as they would before mitosis, but then the cell doesn't divide. This **post-fertilisation polyploidy** becomes important if two closely related species are crossed. The offspring would be **sterile**, because the chromosomes would be **non-homologous** and so couldn't pair up during meiosis. But if the diploid number **doubles**, each chromosome **will** have a homologous one to pair with and meiosis **can** happen. This is thought to have happened to produce the modern **wheat** plant.

Speciation

Darwin's Finches are a good example of Allopatric Speciation

Darwin studied **finches** that live on the Galápagos Islands, a small group of islands 1000 km west of Ecuador, to develop his theory of evolution. He based his theory on his observations:

1) On the Galápagos islands, there are **fourteen** species of **finch** belonging to **four genera**.
2) Each species of finch inhabits a different ecological niche (see p. 48) on the islands and some are only found on one island.
3) The main difference between the finches is the **shape** and **size** of their **beaks**. The birds feed on a variety of different foods from grubs to hard-shelled seeds — each finch has a beak suited to the food it eats.

main food	fruits	insects	insects	cacti	seeds	seeds
feeding adaptation	parrot-like beak	grasping beak	uses cactus spines	large crushing beak	pointed crushing beak	large crushing beak

Despite these differences, Darwin thought that all the finches had a **common ancestor**.
Since then more research has been done which has proved that **geographical isolation** did cause **speciation** on the Galápagos islands. Finches are small birds and it's unusual for them to fly over water, so once a population gets onto an island (perhaps because they were blown off course by a storm) they are effectively **isolated** from the finches on other islands. The differing environmental conditions on each island put **selection pressures** on the birds — and the birds gradually became **adapted** by natural selection to the conditions on the different islands.

Convergent Evolution is when Unrelated Species have Similar Features

Convergent evolution happens when **unrelated** species have **evolved** so that they look very **similar**. For example, **sharks** and **dolphins** look pretty similar and swim in a similar way but they're totally different species — sharks are cartilaginous **fish** and dolphins are **mammals**. They have different evolutionary roots but they have developed similar bone structures to make them well **adapted** for swimming.

Practice Questions

Q1 Define the term 'species'.

Q2 What is a hybrid? Give an example.

Q3 What is the difference between allopatric and sympatric speciation?

Q4 Name four causes of sympatric speciation.

Q5 What was Darwin researching when he proposed his theory of evolution?

Exam Question

Q1 Charles Darwin studied different species of finch on the Galápagos Islands.
a) Describe Darwin's observations. [3 marks]
b) Give an explanation of how Darwin believed the different species developed. [4 marks]

I wish there were biology field trips to the Galápagos Islands...

It's easy to learn the basics of these pages — what a species is and how new ones develop. Then it's just a matter of learning the detail and the correct words for everything. It's important that you know words like 'sympatric' and 'convergent evolution' because they might be used in the exam questions and you'll be stuck if you forget what they mean.

Classification

Classification is all about grouping together organisms that have similar characteristics.
The system of classification in use today was invented by a Swedish botanist, Carolus Linnaeus, in the 1700s.

Classification** is the way Living Organisms are Divided into **Groups

The classification system in use today puts organisms into one of five **kingdoms**:

KINGDOM	EXAMPLES	FEATURES
Prokaryotae	bacteria	unicellular, no nucleus, less than 5 μm, naked DNA in circular strands, cell walls of peptidoglycan
Protoctista	algae, protozoa	eukaryotic cells, usually live in water, unicellular or simple multicellular
Fungi	moulds, yeasts and mushrooms	eukaryotic, heterotrophic, chitin cell wall, saprotrophic
Plantae	mosses, ferns, flowering plants	eukaryotic, multicellular, cell walls made of cellulose, photosynthetic, contain chlorophyll, autotrophic
Animalia	nematodes (roundworms), molluscs, insects, fish, reptiles, birds, mammals	eukaryotic, multicellular, no cell walls, heterotrophic

*All Organisms can be organised into **Taxonomic Groups***

Taxonomy is the branch of science that deals with **classification**.

A **species** is the **smallest** unit of classification (see p. 44-45 for more about species). Closely related species are grouped into **genera** (singular = genus) and closely related genera are grouped into **families**. The system continues like this in a hierarchical pattern until you get to the largest unit of classification, the **kingdom**.

In this system, it's important that there's **no overlap** — a species can only belong to **one** genus, **one** family, **one** order etc.

The Hierarchy of Classification
Kingdom
Phylum
Class
Order
Family
Genus
Species

*For example, **Humans** are **Homo sapiens***

This is how **humans** are classified:

This column shows the features that have been used to classify humans into each of these groups.

		FEATURES
KINGDOM	Animalia	animal
PHYLUM	Chordata	has nerve cord
CLASS	Mammalia	has mammary glands and feeds young on milk, has hair / fur
ORDER	Primates	finger and toe nails, opposable thumb, reduced snout and flattened face, binocular vision, forward-facing eyes
FAMILY	Hominidae	relatively large brain, no tail, skeleton adapted for upright or semi-upright stance
GENUS	*Homo*	cranial capacity > 750 cm^3, upright posture
SPECIES	*sapiens*	erect body carriage, highly developed brain, capacity for abstract reasoning and speech

Classification

The **Binomial System** is used to **Name** organisms

The full name of a human is **Animalia Chordata Mammalia Primate Hominidae *Homo sapiens***. The name gives you a lot of information about how humans have been classified. Using full names is a bit of a mouthful so it's common practice to just give the **genus** and **species** names — that's the **binomial** ('two names') **system**.

The binomial system has a couple of **conventions**:

1) Names are always written in ***italics*** (or they're **<u>underlined</u>** if they're **handwritten**).
2) The **genus** name is always **capitalised** and the **species** name always starts with a **lower case** letter.

e.g.

Human	*Homo sapiens*
Polar bear	*Ursus maritimus*
Sweet pea	*Lathyrus odoratus*

You can also **Classify Organisms** according to how they **Feed**

There are **three** main ways of getting **nutrition** —

1) **Saprotrophic** organisms, e.g. **fungi**, absorb substances from **dead** or **decaying** organisms using **enzymes**.
2) **Autotrophic** organisms, e.g. **plants**, produce their **own** food using **photosynthesis**.
3) **Heterotrophic** organisms, e.g. **animals**, consume complex organic molecules, i.e. they consume **plants** and **animals**.

Practice Questions

Q1 Name the five kingdoms of classification, giving an example organism in each.

Q2 Explain the difference between fungi and plants in terms of how they get their nutrition.

Q3 What are the two rules for using the binomial system?

Q4 What do the phrases saprotrophic, autotrophic and heterotrophic mean?

Exam Questions

Q1 Explain the difference between phylogenetic classification and traditional classification. [2 marks]

Q2 The King Penguin has the scientific name *Aptenodytes patagonicus*. Fill out the missing words a) – e) in the table. [5 marks]

Kingdom	Animalia
a)	Chordata
b)	Aves
Order	Sphenisciformes
c)	Spheniscidae
Genus	d)
Species	e)

Classification — what Vogue magazine did to Wayne Rooney's missus...

The good thing about this is everything is pretty straightforward — don't be put off if lots of the words are new to you (and if 'phylum' is part of your day-to-day vocabulary then I suggest you get out more). You need to learn this thoroughly. In the exam you'll be glad that you did cos there's often some easy marks to be had about this kind of stuff.

Ecosystems and Energy Transfer

These two pages deal with loads of words that you need to know. They also deal with the way energy flows through ecosystems. Not too difficult, but you need to understand the basic principles.

You need to learn some **Definitions** to get you started

Ecosystem	An **ecological unit** which includes all the **organisms** living in a particular area and all the **abiotic** (non-living) features of the local environment.
Population	All the **individuals** of a particular **species** living in a given area.
Community	All the **living organisms** in an ecosystem. These organisms are all **interconnected** by food chains and food webs.
Habitat	The **place** where the communities live, e.g. a rocky shore, a field, etc.
Niche	The **'role'** an organism has in its environment — where it lives, what it eats, where and when it feeds, when it is active etc. Every species has its own **unique** niche.
Environment	The **conditions** surrounding an organism, including both abiotic factors (e.g. temperature, rainfall) and biotic factors (e.g. predation, competition).

The opposite of abiotic is biotic (to do with living things).

Don't get confused between population size (how many in total), and density (how many in a given area).

Energy **Flows Through** ecosystems

Energy comes into the ecosystem from sunlight and is fixed into the ecosystem by plants during **photosynthesis**. The energy stored in the plants can then be passed onto other organisms in the ecosystem along **food chains** — each link in a food chain is called a **trophic level**. During this process a lot of the energy is gradually lost from the food chain — this is **dissipation**. **All** the food chains in an **ecosystem** are linked together in **food webs**.

Scientists call food chains and webs dynamic feeding relationships.

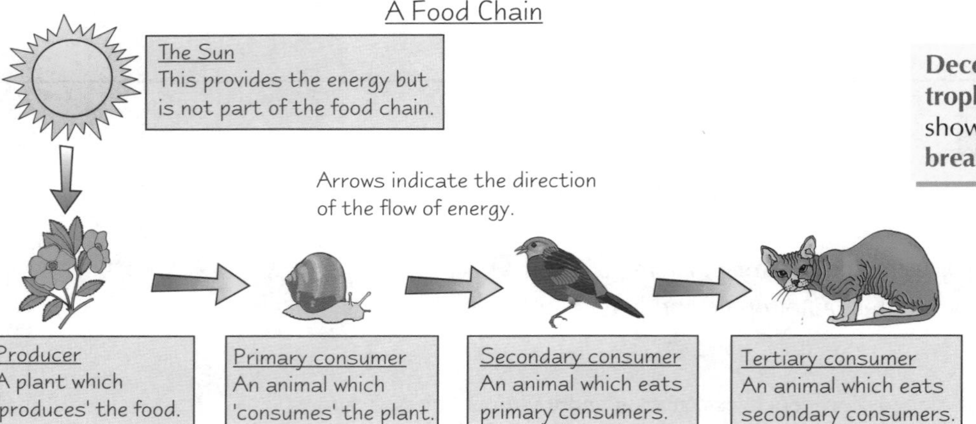

Decomposers are another **trophic level**, (they're not shown on this diagram). They **break down** dead material.

1) At each trophic level about **10%** of the energy is used for **growth** and **storage** — that's the energy that can be passed onto the **next** level when the organism is **consumed**.
2) So about **90%** of energy is **wasted** between one trophic level and the next. The cow diagram shows what typically happens to this energy.

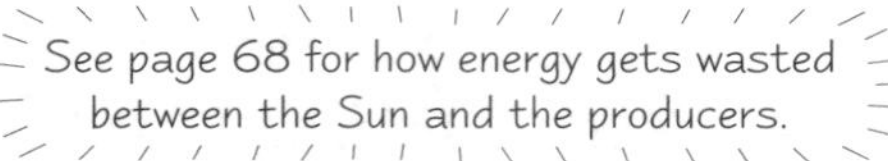
See page 68 for how energy gets wasted between the Sun and the producers.

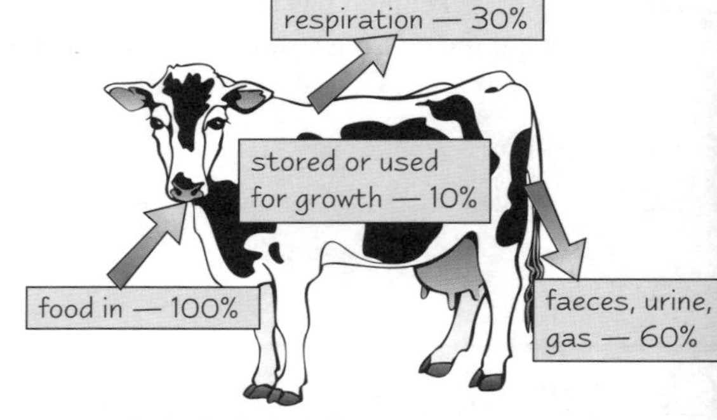

All this wastage means that food chains rarely get longer than five links.

Ecosystems and Energy Transfer

Food chains can be shown as **Three Types** of **Pyramid**

In all food pyramids the **area** of each block tells you about the **size** of the trophic level.
The food chain can be shown in terms of number, biomass or energy:

Pyramids of Number

These are the **easiest to produce** — they show the **numbers** of the different organisms so it's just a question of counting. They're **sometimes misleading**, though — the nice pyramid shape is often messed up by the presence of small numbers of big organisms (like trees) or large numbers of small organisms (like parasites).

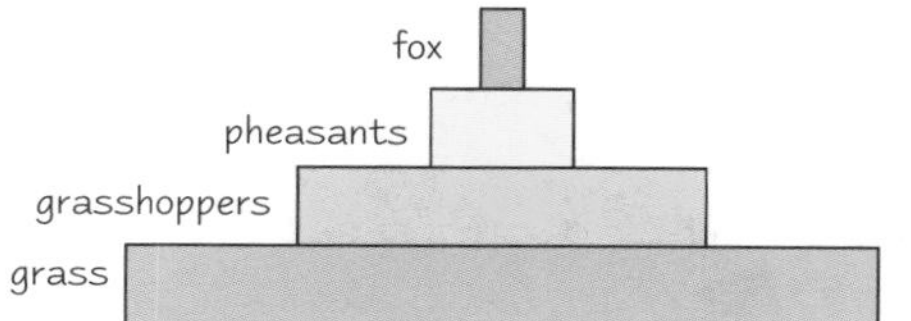

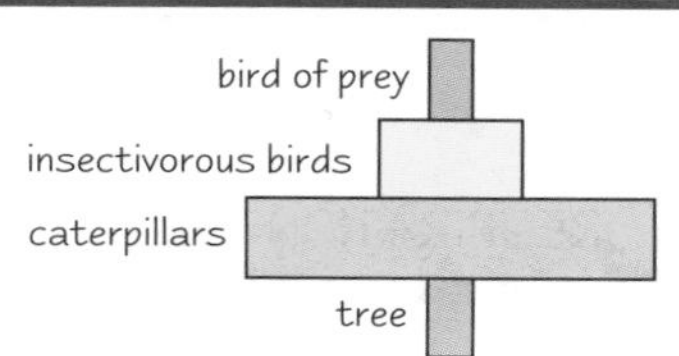

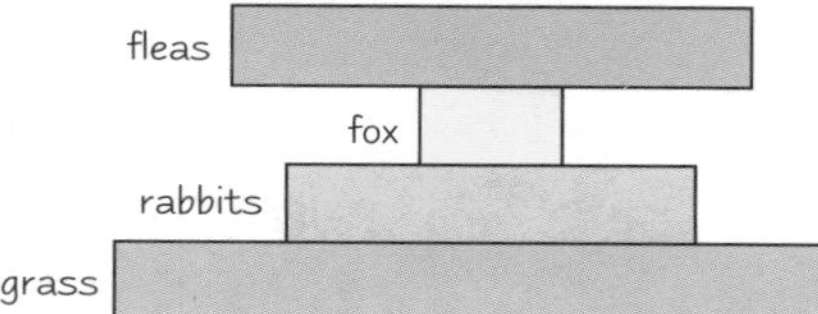

Pyramids of Biomass

These are produced by measuring biomass (the **dry mass** of the organisms in kg/m^2).
It's **difficult to get the raw data** for them (you'd have to kill the organisms) but they're pretty accurate — they nearly always come out pyramid-shaped.

Pyramids like this are always symmetrical and they're always drawn to scale.

bird of prey
insectivorous birds
caterpillars
tree

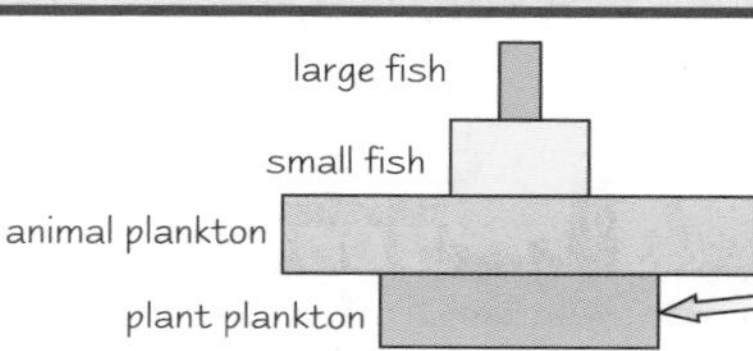

The amount of plant plankton is quite small at any given instant, but, because it has a short life and reproduces very quickly, there is a lot of it around over a period of time. This is why the plant plankton level is smaller than the animal plankton level.

Pyramids of Energy

These measure the **amount of energy** in the organisms in **kilojoules** per **square metre** per **year** ($kJm^{-2}yr^{-1}$).
This data is **very difficult to measure** but these pyramids give the **best picture of the food chain** and are **always proper pyramids**.

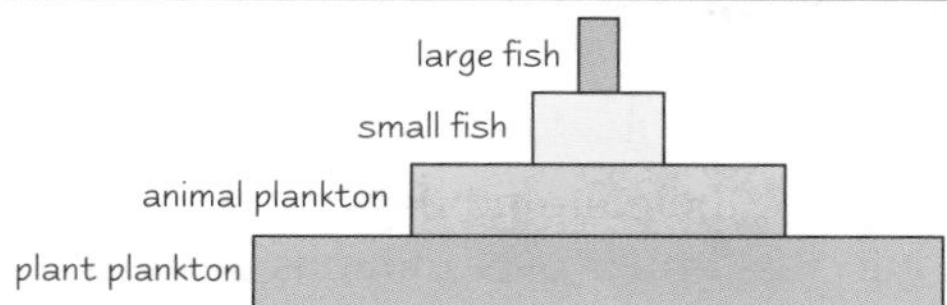

Practice Questions

Q1 What is a community?

Q2 What is a trophic level?

Q3 Why are there very rarely more than five links in a food chain?

Q4 What are the three types of pyramid?

Exam Questions

Q1 Explain the differences between a habitat and an ecosystem. [4 marks]

Q2 Pyramids of number and pyramids of biomass are not always pyramid-shaped, but pyramids of energy are. Explain why this is so. [4 marks]

Ah... pretty coloured pyramids — for A2 Biology, this is pure heaven...

This stuff is fairly straightforward, but there are quite a few definitions you need to get in your head before the rest of the section. Otherwise, come the harder stuff, you'll be struggling to remember what ecosystems, populations, communities and habitats are. Make sure you know how energy flows through ecosystems and the different types of pyramid as well.

Nutrient Cycles

The amount of carbon and nitrogen on Earth is fixed — they can exist in different forms but no more can be made. The good news is they are constantly cycled around so they won't run out. The bad news is that you have to learn how that happens — just like you did for GCSE — only this time it's a bit more complicated. Great.

*The **Carbon Cycle** is fairly straightforward*

The carbon cycle involves four basic processes – **photosynthesis**, **respiration**, **death and decay** and **combustion**.

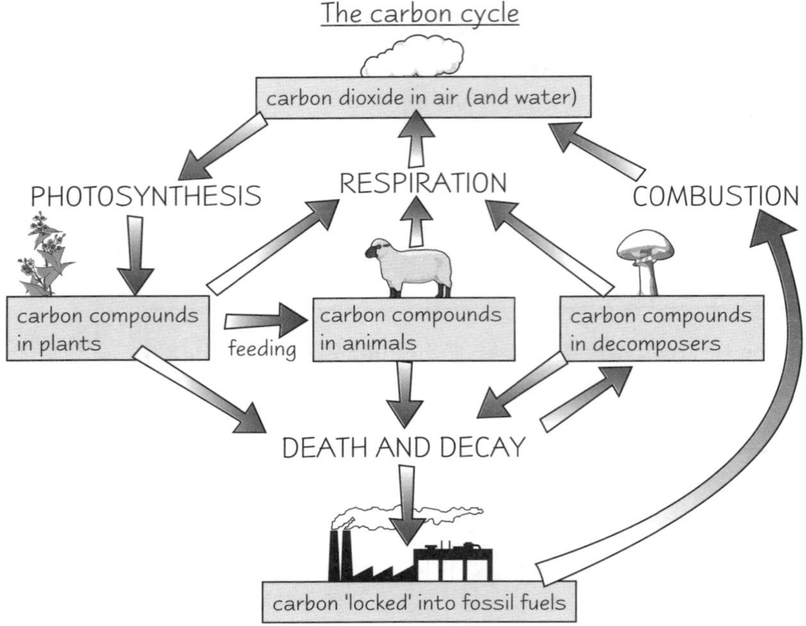

These are the seven things that you need to **remember** about the carbon cycle:

1) The only way that carbon gets into ecosystems is through **photosynthesis**.
2) Herbivores get their carbon by eating **plants**, carnivores get theirs by eating **other animals** and omnivores eat a **mixture** of plants and animals.
3) **Decomposers** get their carbon by digesting dead organisms. Feeding on dead material is **saprobiontic** or **saprotrophic nutrition**.
4) All living organisms return carbon to the air in the form of **carbon dioxide** through **respiration**.
5) If plants or animals **die** in situations where there are no decomposers (e.g. deep oceans) the carbon in them can get turned into **fossil fuels** over millions of years.
6) The carbon in fossil fuels is released when they are burned — **combustion**.
7) **Microorganisms** are important in the cycle because they can quickly get the carbon in dead material **back into the atmosphere**.

*The **Carbon Cycle** keeps atmospheric carbon dioxide levels **Constant***

In a totally **natural** situation the carbon cycle would keep atmospheric levels of carbon dioxide **more or less the same**.
Nowadays people are affecting the **global carbon balance** in two key ways:

- We burn huge quantities of **wood** and **fossil fuels** each year, which **adds** loads of carbon dioxide to the air.
- We are **clearing large areas of forest**, which would normally help to absorb some of the carbon dioxide in the atmosphere.
- In combination these two activities mean that the total amount of carbon in the atmosphere is much **higher** than it naturally would be.

Nutrient Cycles

*The **Nitrogen Cycle** is a bit more **Complicated***

You need to be familiar with all the stages in the nitrogen cycle:

Don't worry if you get a diagram that looks different to this in the exam - all the information will be basically the same.

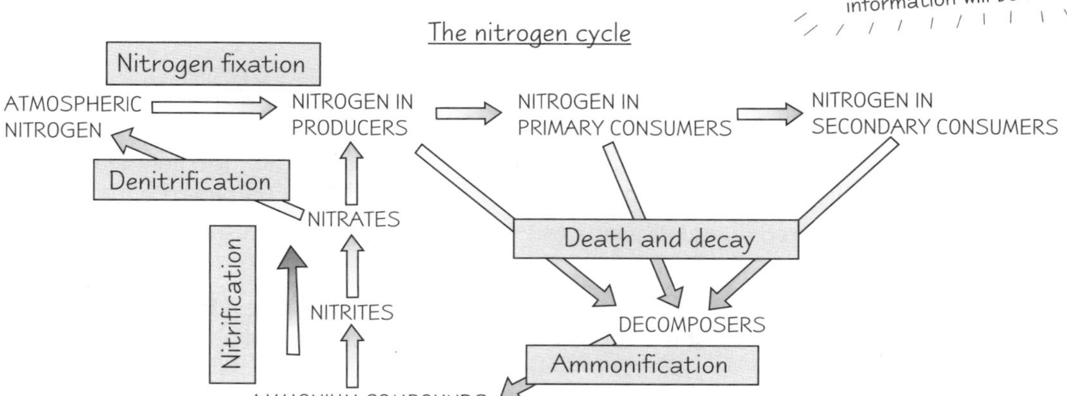

Plants and animals need nitrogen for **proteins** and for nucleic acids, but despite the atmosphere being 78% nitrogen, neither of them can use nitrogen gas. The key thing you need to remember is how important **bacteria** are (see box below). Without bacteria to produce the **nitrates** that plant roots can absorb, plants and animals couldn't exist.

1) Atmospheric nitrogen is **fixed** by **bacteria**. Some **live free** in the soil, others are found inside root nodules of **leguminous** plants (that's peas, beans and clover to you and me). Atmospheric nitrogen is changed into **ammonia**, and then into amino acids by the bacteria.
2) The nitrogen in the plant proteins is passed onto animals through **food chains**.
3) When living organisms **die** their nitrogen is **returned** to the soil in the form of **ammonium compounds** by **microorganisms**. Animals get rid of excess amino acids via **deamination** in their livers — the nitrogen gets back into the soil via their **urine**.
4) Ammonium compounds are changed into nitrates by **nitrifying bacteria**. Firstly *Nitrosomonas* changes ammonium compounds into nitrites, then *Nitrobacter* changes the nitrites into nitrates.
5) Nitrates are **converted back** into atmospheric nitrogen by **denitrifying bacteria**.

Don't worry — you don't need to learn the names of the microorganisms.

Sometimes you'll see a couple of extra things on diagrams of the nitrogen cycle:

1) **Industrial processes** like the Haber Process produce ammonia and nitrate fertilisers directly from atmospheric nitrogen.
2) **Lightning** naturally converts nitrogen into nitrates.

Practice Questions

Q1 What process in living things extracts carbon dioxide from the air?

Q2 What is 'saprobiontic nutrition'?

Q3 What types of plants have root nodules?

Q4 In the nitrogen cycle, what chemical changes occur during 'nitrification'?

Exam Questions

Q1 Explain how the carbon cycle has maintained the level of carbon dioxide in the atmosphere and how human activity has disrupted this balance. [10 marks]

Q2 Describe the role and significance of microbes in the nitrogen cycle. [6 marks]

The Carbon and Nitrogen Cycles — surely, not again...

Here we are in A2 Biology and the carbon and nitrogen cycles are back again. When you realise that without nitrogen-recycling bacteria, plants and animals couldn't exist, then you can see how important the cycles are. Perhaps that's why they keep cropping up, or perhaps the examiners are just torturing you — either way it's got to be learnt.

Studying Ecosystems

If you've been on an ecology field trip you'll be familiar with this stuff. You'll be relieved to know that you can revise this in the comfort of your own bedroom — you won't be asked to stand in a river catching horrible squirmy things.

You need to know how to take ***Abiotic Measurements***

Temperature is easy enough — just use a **thermometer**.

pH measurements are only taken for soil or water. **Indicator** paper / liquid or an electronic **pH monitor** are used to get the data.

Light intensity is difficult to measure because it varies a lot over short periods of time. You get the most accurate results if you connect a **light sensor** to a data logger and take readings over a period of time.

Oxygen level only needs to be measured in aquatic habitats. An **oxygen electrode** is used to take readings.

Air humidity is measured with a **hygrometer**.

Moisture content of soil is calculated by finding the mass of a soil sample and putting it in an oven to dry out. The amount of mass that has been lost is worked out as a % of the original mass.

Quadrat Frames *are a basic tool for* ***Ecological Sampling***

Ecologists look at three key **factors** when they're working out diversities:

SPECIES FREQUENCY	This is how **abundant** a species is in an area.
SPECIES RICHNESS	This is the **total number** of **different species** in an area.
PERCENTAGE COVER	This is how much of the surface is covered by a particular plant species (you can't use it for **animals** because they move around too much).

1) To measure all of these you use a piece of equipment called a **frame quadrat** — a square frame made from metal or wood. The area inside this square is known as a **quadrat**.

A frame quadrat

quadrat frame

the area of this quadrat is $0.25m^2$

50cm

50cm

2) **Quadrat frames** are **laid on the ground** (or the river / sea / pond bed if it's an aquatic environment). The **total number** of **species** in the quadrat frame is recorded as well as the number of **individuals** of each species.

3) Generally it's not practical to collect data for a whole area (it would take you ages) so **samples** are taken instead. This involves sampling from lots of quadrats from different parts of the area. The data from the samples is then used to **calculate** the figures for the **entire area** being studied. **Random sampling** is used to make sure that there isn't any **bias** in the data.

4) **Species frequency** is measured by counting **how many quadrats** each species appears in and is given as a percentage (e.g. if a species was found in 5 out of 20 quadrat samples, the frequency would be 25%).

5) **Species richness** is measured by counting up the **total number of species** found in all the samples. You assume that the number of different species in your sample is the same as the number in the whole area that you are studying.

6) **Percentage cover** is measured by dividing the area inside the quadrat frame into a **10 × 10 grid** and counting **how many squares** each species takes up. Sometimes plants **overlap** so the total percentage cover ends up being **more** than **100%**.

Measuring % cover

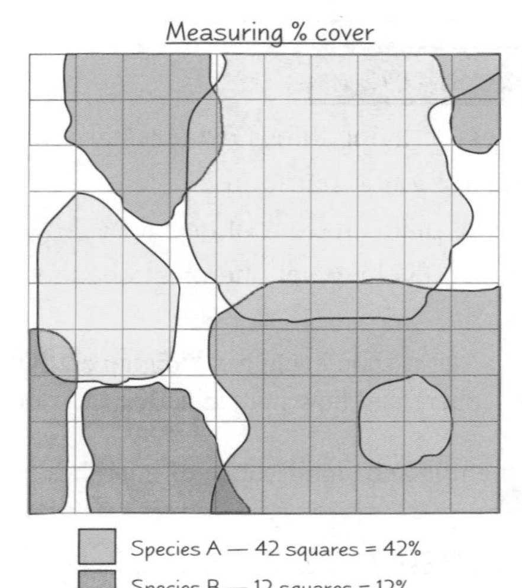

You count a square if it's more than half-covered.

Studying Ecosystems

Think about the Size and Number of Quadrats to sample

When using quadrat frames, remember that samples must be taken **randomly** (see page 54). It's also important to consider the **size** of the quadrat — smaller quadrats give more accurate results, but it takes longer to collect the data and they're not appropriate for large plants and trees.

Plotting a **graph** of **cumulative** number of species found against number of quadrats sampled should show you how many quadrats you'd need to sample in further studies of the same type of habitat.

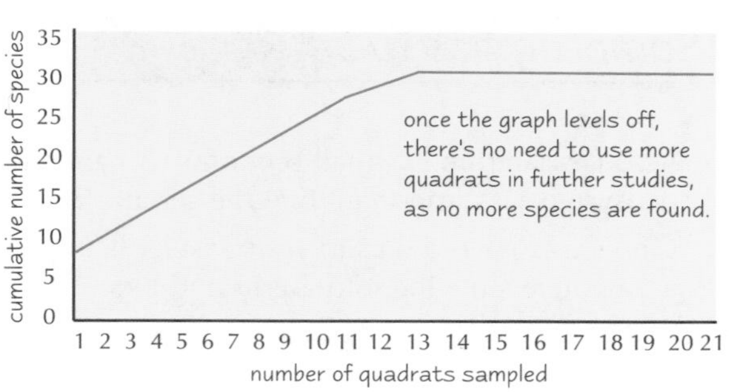

Point Quadrats are an alternative to quadrat frames

Pins are dropped through holes in the frame and every plant that each pin hits is recorded. If a pin hits several **overlapping** plants, **all** of them are recorded. A tape measure is laid along the area you want to study and the quadrat is placed at regular intervals (e.g. every 2 metres) at a right angle to the tape.

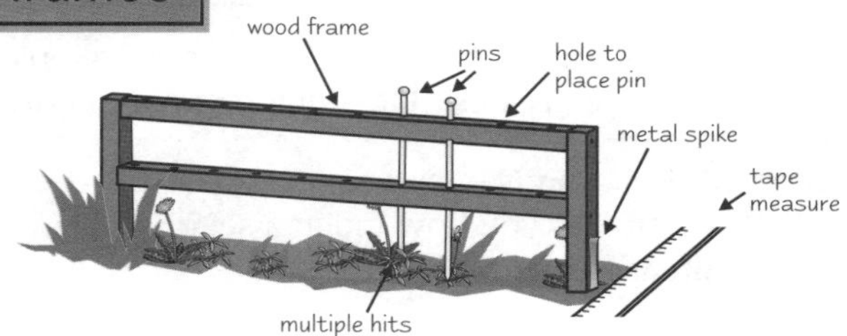

Line and Belt Transects are used to Survey an area

The line you select to sample across the area is called a **transect**. Transects are useful when you want to look for **trends** in an area e.g. the **distribution of species** from low tide to the top of a rocky shore.

A **line transect** is when you place a tape measure along the transect and record what species are touching the tape measure.

A **belt transect** is when data is collected between two transects a short distance apart. This is done by placing frame quadrats next to each other along the transect.

If it would take ages to count all the species along the transect, you can take measurements at set intervals, e.g. 1 m apart. This is called an **interrupted transect**.

The data collected from belt or line transects is plotted on a **kite diagram** (that's just a fancy kind of graph) and trends across the area can be observed.

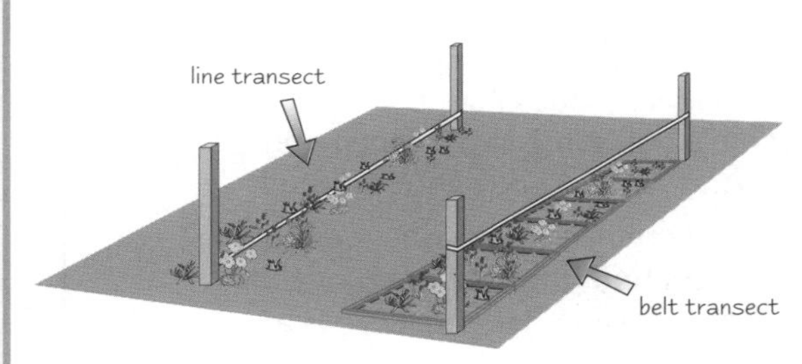

Practice Questions

Q1 What three factors do ecologists look at when they're working out diversities?

Q2 Briefly describe what you do with a frame quadrat.

Q3 What is the difference between a line transect and a belt transect?

Exam Questions

Q1 Under what circumstances would you use a transect rather than random sampling of an ecosystem? Give an example in your answer. [2 marks]

Q2 Describe in detail how you would measure the percentage cover of clover on a school field. [8 marks]

It's just all a bit random...

Well actually, that's the point, it's supposed to be random, or it wouldn't be a true representation. If you've used a quadrat, you've probably already realised it's just a fancy name for four bits of wood nailed together. They're probably expensive too. If you're a business studies student, you might see a money-making opportunity here, but get lost, because I saw it first.

Studying Ecosystems

Sheets and sheets of muddy paper with your field results don't tell you much about an ecosystem. To work out what they really tell you, you need to do some sums...

Sampling has to be **Random**

If you're sampling a small section of an area and then drawing conclusions about the whole ecosystem, it's important that the sample **accurately** represents the ecosystem **as a whole**.

One way to avoid bias in your answer is to pick the sample sites **randomly**, e.g. you could divide the whole area you're studying into a **grid** and then use a random number generator or table to select each coordinate.

Once you have your data you've got to **Analyse It**

In a **normal distribution** of data, most of the samples are **close to the mean** (the average value), with relatively **few** samples at the **extremes**. On a graph, a normal distribution produces a **bell-shaped curve**, like the one below.

Standard deviation is often used to analyse data sets — it tells you how much a set of data is **spread out** around the **mean**.

E.g. you might use standard deviation to find the variation in the number of apples produced by trees in an apple orchard.

The formula for standard deviation (s) is:

$$s = \sqrt{\frac{\sum x^2 - \frac{(\sum x)^2}{n}}{n-1}}$$

s is the standard deviation
Σ means 'sum of'
x is an individual result
n is the total no. of results

When all the samples have **similar** values then the distribution curve is **steep** and the standard deviation is **small**.

When the samples show a **lot** of **variation**, the distribution curve is relatively **flat** and there is a **large** standard deviation.

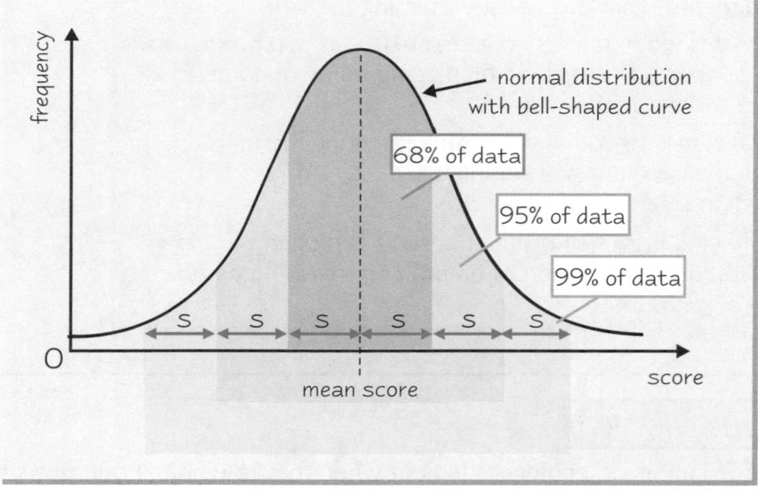

You can use the standard deviation to split any normal distribution graph into the areas shown:

- The middle area (within **1 standard deviation** either side of the mean) shows you where **68%** of the data lies.
- The next area (within **2 standard deviations** either side of the mean) shows you where **95%** of the data lies.
- The biggest area (within **3 standard deviations** either side of the mean) shows you where **99%** of the data lies.

These numbers are always the same.

Studying Ecosystems

The **Chi-Squared** test is really **Useful** in ecology

The chi-squared test is used to test a **null hypothesis**. This will assume that there's **no statistically significant difference between sets of results**. The chi-squared test checks whether any **difference** between the expected and the observed results is **significant** or not.

For example, you could see whether there are any significant differences between the numbers of salmon caught in different rivers. The **null hypothesis** is that the **catch** will be the **same** in all rivers (I know you wouldn't *really* expect that, but it's an assumption of the test).

The **formula** you use to work out chi-squared is:

$$x^2 = \sum \frac{(O-E)^2}{E}$$

Where:
O = observed result
E = expected (null) result

You can only use your actual numbers when working out your result, not percentages. Oh, and don't panic — that equation will be provided in the exam.

The value you get is then looked up in a **chi-squared table** like the one on the right. You need to know the **degrees of freedom** to look up your result. This is just the number of categories (classes) **minus 1**. In the salmon experiment, four different rivers are sampled, so there are **four** categories. This means the degrees of freedom are simply 4 – 1 = **3**.

degrees of freedom	no. of classes	X^2 values					
1	2	0.46	1.64	2.71	3.84	6.64	10.83
2	3	1.39	3.22	4.61	5.99	9.21	13.82
3	4	2.37	4.64	6.25	7.82	11.34	16.27
4	5	3.36	5.99	7.78	9.49	13.28	18.47
probability that deviation is due to chance alone		0.50 (50%)	0.20 (20%)	0.10 (10%)	0.05 (5%)	0.01 (1%)	0.001 (0.1%)

In the salmon example, the **observed results** were:

	River A	River B	River C	River D
Number of salmon	22	18	9	25

The **expected result** is basically a **mean** of the observed results. This is because it's based on the null hypothesis, which predicts no difference between the data sets. To calculate the expected results, **add together** the observed results and then **divide** by the number of data sets.

E = 22 + 18 + 9 + 25 ÷ 4 = **18.5**

River	O	E	(O-E)	$(O-E)^2$	$(O-E)^2 \div E$
A	22	18.5	3.5	12.25	0.66
B	18	18.5	-0.5	0.25	0.01
C	9	18.5	-9.5	90.25	4.88
D	25	18.5	6.5	42.25	2.28

Calculate $(O-E)^2 \div E$ for each data set (i.e. for each river). Then **add them all together**:

0.66 + 0.01 + 4.88 + 2.28 = **7.83**

Looking this final number up in the chi-squared table shows that it corresponds to a probability of between **0.05** and **0.01** that the deviation is due to **chance**.

In the chi-squared test, the **critical value** is **p = 0.05**. If the probability that the deviation is due to chance is **more** than p = 0.05 (i.e. a probability of more than **5%**), we can **accept the null hypothesis** that there is **no** statistically significant difference between the observed and the expected results.

In the case of our salmon, however, the probability is a bit **less** than 5%. So you have to **reject** the null hypothesis and say that there **is** a **statistically significant difference** between the results you got and the results you'd expected to get. Some **other factor** must be affecting the results, because there's a significant difference in salmon catches between rivers.

Practice Questions

Q1 What does standard deviation give an indication of?

Q2 What is a null hypothesis?

Q3 What is the critical value in the chi-squared test?

Exam Question

Q1 A scientist investigated the effect of fertiliser on the yield of a turnip crop. She compared two crops of turnips, one with fertiliser, and one without. She found the mean yields of both groups, did a chi-squared test and found a chi-squared value of 2.53. Does fertiliser make a significant difference? Explain your answer [use the chi-squared table on this page]. [3 marks]

Difficulty with statistics is a standard error...

The important thing is being able to interpret results. That means that you need to be able to tell the examiner if things are significant or not, and you can tell that a large standard deviation means there's lots of variation. Make sure that you're confident with the chi-squared test — a little practice would help a lot. Then run away and hide under a bush somewhere.

Dynamics of Ecosystems

Here's a lovely easy couple of pages for you. Some things eat each other, some things compete for food, populations get bigger, and they sometimes get smaller too. That's about it really. A lovely breather after the chi-squared test.

Population Growth is stopped by Limiting Factors

Limiting factors are things that put an **upper limit** on a population's size.

Many limiting factors are **abiotic** (non-living) — examples include: **temperature**, **soil** or **water pH**, **oxygen availability**, and the supply of **mineral nutrients**.

Some of these things (e.g. temperature and oxygen availability) can affect the **rate** of **population growth** as well as limiting the total **size**.

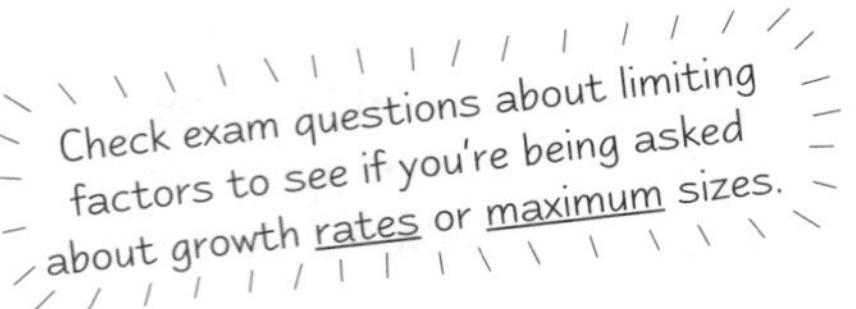

Interactions between Organisms can Limit population size

There are two important interactions that limit population size — **competition** and **predation**.

1) Competition occurs when a lot of organisms are competing for some sort of limited **'resource'** — very often it's food, but it can be other things like **shelter**, **nesting sites** and **mates**.
2) If the organisms competing are of the **same** species, it is called **intraspecific** competition. If they are from **different** species, it is called **interspecific** competition.

Interspecific Competition affects population Distribution

Sometimes interspecific competition doesn't just affect the population **size** of a species, it totally **prevents** the species living in an area at all.

Since the introduction of the **grey squirrel** into Britain, the native **red squirrel** has **disappeared** from large areas because of interspecific competition. In the few areas where both species still live, both populations are **smaller** than they would be if there was only one kind of squirrel there.

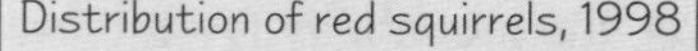
Distribution of red squirrels, 1998

Distribution of grey squirrels, 1998

Intraspecific Competition occurs Within a species

Intraspecific competition happens between members of the **same species** in a population. As the population grows, there will be more competition for **space** and **food**.

The population tends to **grow** until some factor becomes **limiting**, and then it will begin to **decline**. A **smaller** population then means that there's **less** competition for space and food, which is **better** for survival and reproduction — so the population starts to **grow** again.

The population varies around an **average** figure, which is known as the **carrying capacity**.

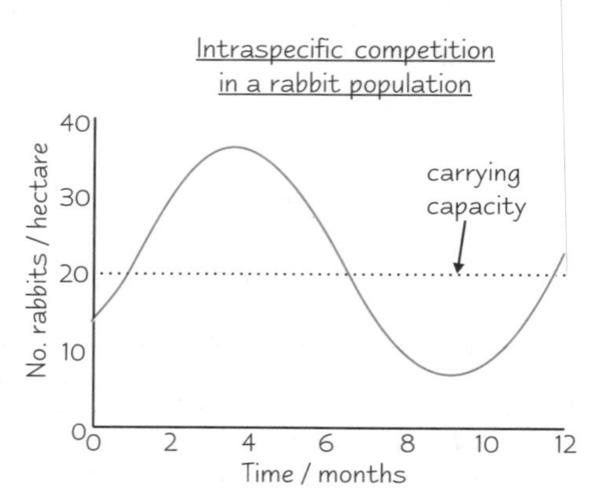

Dynamics of Ecosystems

Predator and **Prey** populations are interlinked

The presence of **predators** affects the **size** of populations. The **graph** on the right shows **population fluctuations** of the **lynx** and its prey, the **snowshoe hare**, in Canada — it's a good example of how closely related predator and prey numbers can be.

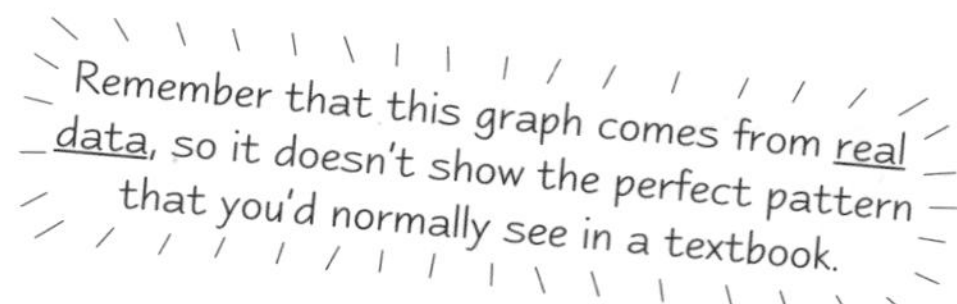

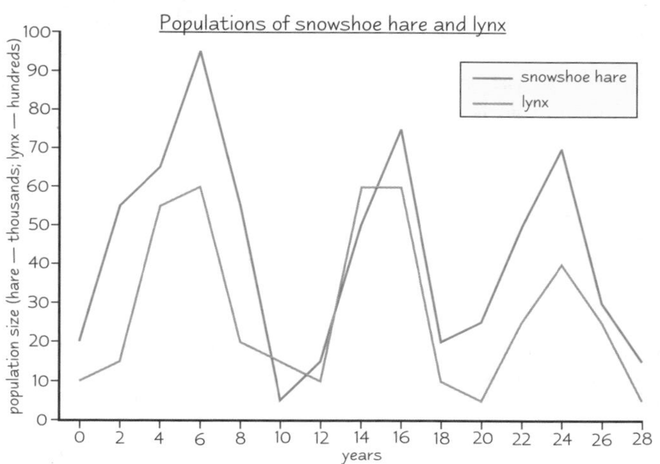

1) The theory is that as the **population** of **prey grows**, there is **more food for predators**, and so the predator **population grows**.
2) As the **predator** population gets **larger**, **more prey** is **eaten** so the **prey** population **falls**.
3) Then, there is **less food** for the **predators** so their **numbers go down**, and so on.
4) In nature you rarely get this **exact pattern** because it's unusual to get a situation where a species is eaten by just **one** predator, or a predator just eats **one** prey species.

Some regulating factors are '**Density Independent**'

Most of the factors which limit populations mentioned so far, like food, shelter, mates etc. are **density dependent** — in other words, they have a greater impact as a population gets more **dense**. **Disease** is another density dependent factor.

Some other factors like **flooding** and **forest fires** are **density independent** — they can affect populations of any density. These factors are completely **unrelated** to **population density**. Most **abiotic** factors are density independent. See page 62 for more on density dependent and density independent factors.

Practice Questions

Q1 State three abiotic factors that might limit the size of a population.

Q2 Explain the difference between interspecific and intraspecific competition.

Q3 Give an example of an animal in which interspecific competition has affected population distribution.

Q4 What will happen to the size of a population when competition between its members is low?

Q5 What does the term 'carrying capacity' mean?

Q6 What does the term 'density dependent' mean?

Exam Question

Q1 Factors that regulate population size can be density dependent or density independent. Explain the difference between these two types. [4 marks]

Ecosystems — not quite as dynamic as I expected...

Again there's some stuff you've seen lots before here — like predator-prey relationships, but it's jumbled up with new stuff too. So don't be fooled into thinking, "yeah yeah, been there, done that, got the T-shirt — in fact I invented the T-shirt me, this A2 biology lark's a cinch, reckon I might just sack this revision off"... you get the point, this stuff needs learning.

Succession

Succession is all about how ecosystems change over time. Apart from a few fancy words I think that it's one of the easiest things in A2 biology — it's a lot more straightforward than the Krebs cycle and photosynthesis.

There are **Two** different types of **Succession**

Succession is sometimes called ecological succession in exams.

Succession is the process where **plant communities** gradually develop on **bare land**.
Eventually a **stable climax community** develops and after that big changes don't tend to happen.

There are **two** different types of succession...

Primary succession

Happens on land where there is no proper soil and **no living organisms**. New land created by a **volcanic eruption** is a good example of a place where primary succession will occur.

Secondary succession

Happens when most of the living organisms in an area are **destroyed**, but the **soil** and **some** living organisms remain. Examples include: woodland that has been burned by a **forest fire**, areas subject to severe **pollution**, or land that is cleared by **people** for things like **housing** or **new roads**.

Each stage in the succession of an area is called a **seral stage**. In every seral stage the plants change the environmental conditions, making them suitable for the next plants to move in.

You need to **Learn** an **Example** of succession

This example shows the seral stages that change **bare** sand dunes into **mature woodland**.

You need to learn this example, including the names of each of the communities.

The first plants to colonise an area have to be <u>specialised</u> so they can deal with the <u>harsh abiotic conditions</u>. These plants are known as <u>pioneer species</u>. They are usually herbaceous (non-woody).

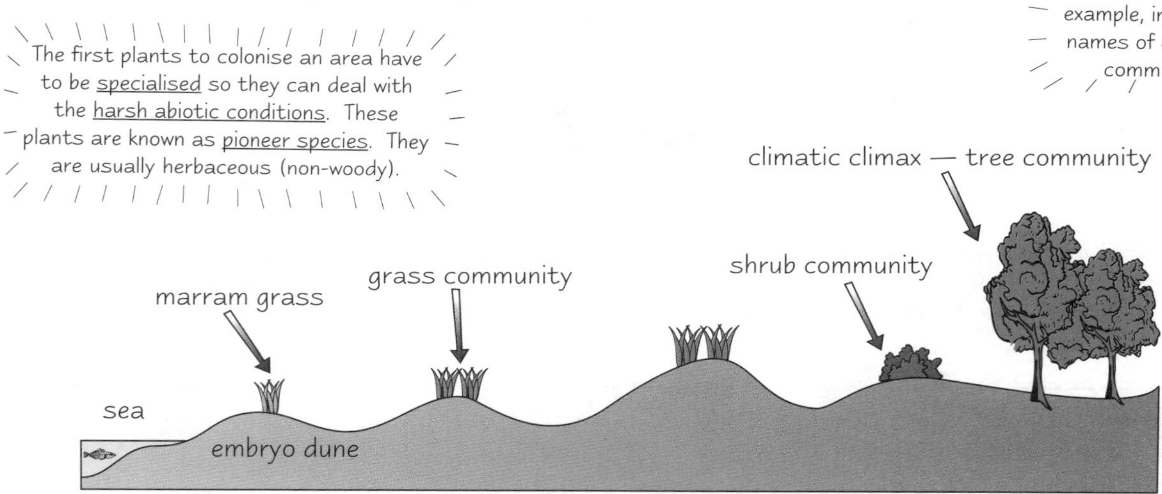

1) The first **'pioneer'** species to colonise the area need to be able to cope with the **harsh abiotic conditions** on the **sand dunes** — there is **little fresh water** available, there are **high salt levels**, the **winds** are **strong** and there is **no proper soil**. Marram grass has good **xerophytic** adaptations (see page 66) so it is usually the first to start growing.
2) As the pioneer species begin to **die** they are broken down by microorganisms. The dead marram grass adds **organic material** to the sand creating a very basic **'soil'** which can hold more water than plain sand.
3) This soil means that the abiotic conditions are **less hostile** and so other, less specialised grasses begin to grow.
4) These new grasses will eventually **out-compete** the original colonisers via **interspecific** competition.
5) As each new species moves in, more **niches** are created making the area suitable for even more species.
6) After the grass communities have all been out-competed, the area will be colonised by **shrubs** like **brambles**.
7) Eventually the area becomes dominated by **trees** — in Europe the trees will usually be things like **birch** and **oak**. The trees dominate because they prevent light from reaching the herbaceous plants below the leaf canopy.

Succession

Diversity Increases and *Species Change* as succession progresses

When succession happens in any environment the general pattern of change is always the same:

- The species present become **more complex** e.g. a forest starts with simple mosses and finishes with trees.
- The **total number** of organisms **increases**.
- The **number of species increases**.
- **Larger species** of plants arrive.
- **Animals** begin to move into the area — with each seral stage **larger** animals move in.
- **Food webs** become more **complex**.
- Overall, these changes mean that the ecosystem becomes more **stable**.

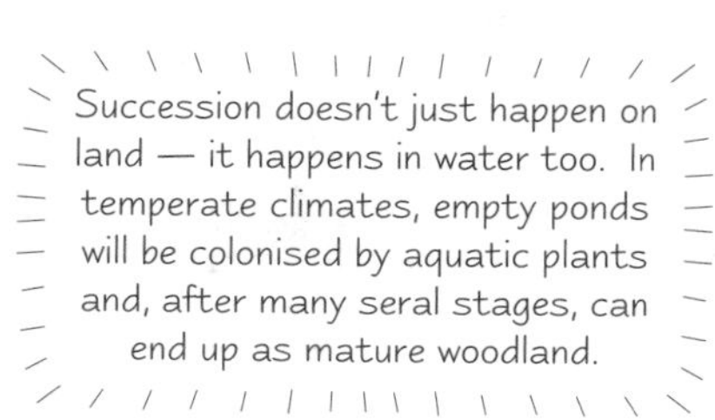

There are different *Types* of *Climax Community*

Various factors can **stop succession** going any further and lead to a **climax** community.
The climax is **classified** according to **what** has prevented the succession from going any further...

1) In a **climatic climax**, the succession has gone as far as the **climate** in the area will allow. E.g. Trees can't grow at high altitudes, so high up on alpine mountains the largest plants are **shrubs**.
2) **Human activities** can stop succession by felling trees, ploughing fields or grazing animals on farmland. Some ecosystems are deliberately '**managed**' to keep them in a particular state, for example, the heather on moorland is burned every 5-7 years to prevent woodland from developing. When succession is stopped **artificially** like this the climax community is called a **plagioclimax**.

Farmland is an example of a *Plagioclimax*

Succession is stopped by regular **ploughing** or by the **grazing** of stock. In a grazed field, **grass** can survive because it is fast-growing, but slow-growing plants get eaten before they can get established.

If the **grazing stops**, then slower-growing plants can gradually begin to establish themselves. As they do, the **grass** will become **less dominant**. The new plant species will attract a wider range of **insects** and so the area will **increase** its **diversity**. Eventually, the field will become replaced by **woodland**.

Practice Questions

Q1 What is secondary succession?

Q2 What are the stages in succession called?

Q3 What is a pioneer species?

Q4 What name is given to a climax community brought about by human intervention?

Exam Questions

Q1 Define the term 'ecological succession', explaining how it occurs and the different types of climax communities that can be produced. [8 marks]

Q2 a) Suggest three features of a plant species that might make it a successful pioneer species. [3 marks]
b) Suggest two reasons why such a pioneer species may disappear during the early stages of succession. [2 marks]

There are many different types of climax...

If your enthusiasm for all these Biology facts is waning, why not try reading some ICT... "Remote management supports users on a network. In the event of a significant problem or recurring error, the network administrator takes temporary control. Afterwards, the user can be advised on how to avoid similar problems in the future..." You see what I'm saying?

Agriculture and Ecosystems

Agriculture has always changed ecosystems.
You need to know the facts and have a balanced view on modern and organic farming methods.

Farming can be **Intensive** or **Extensive**

The **traditional** methods of farming and those used by **organic** farmers are examples of **extensive food production**. In the later part of the twentieth century, these methods were replaced by **intensive farming**. The table summarises the differences:

Intensive	Extensive
Requires a lot of capital investment	Less capital investment needed
Requires less space	Requires a lot of space to grow the same amount of produce
Not labour intensive — much is done by machines	Labour intensive
Production concentrated on single crop (monoculture) or animal species	Farm usually grows several crops and/or produces livestock
Large amounts of pesticides used Biological control sometimes used	Fewer or no pesticides used Biological control more common
Large scale use of inorganic fertilisers	Emphasis on organic fertilisers

Intensive Farming methods can affect the **Environment**

Intensive farming methods have **boosted food production**, stopping people starving in some parts of the world and making sure there is always plenty of good quality food on supermarket shelves.

But these methods can cause **environmental problems**:

- **Hedgerows** are **removed** to allow access for large machines or to enlarge fields. This **destroys** the **habitat** of many species and reduces species diversity.
- Large amounts of **inorganic fertiliser** are used. This can **leach** through the soil and **pollute** nearby streams and rivers (see below).
- '**Factory farming**' methods of keeping many animals in a small space produce large amounts of **waste**, which can cause pollution if not disposed of carefully.
- Large scale use of **pesticides** can result in water **pollution** and may also kill non-pest species. Food can become contaminated.

The use of **Fertilisers** can lead to **Eutrophication**

Fertilisers come in **two types** — **inorganic** (man-made chemicals) and **organic** (manure).

1) **Organic** fertiliser is better for the **environment**, but is **more difficult** to **store** and **apply**. Its bulk improves the soil structure.

2) Fertilisers often get **washed** into nearby streams and rivers, which is a **waste**, and can also cause **eutrophication**.

Eutrophication occurs when **fertilisers** stimulate the **growth** of **algae** in ponds and lakes. These algae prevent light from reaching the water plants below.
Eventually the plants **die** because they are unable to photosynthesise enough. The dead plants are decomposed by **bacteria**. The bacteria drastically **reduce** the **oxygen** levels in the water, which kills fish and other aquatic organisms.
Phosphates entering water from **sewage** also have this effect.

Digging up farmland that hasn't been used for a while releases nitrates which can cause eutrophication. Nitrates from soil can also contaminate drinking water.

Agriculture and Ecosystems

Farmers can be Environmentally Friendly

Organic farmers set out to be 'green' and avoid all use of chemicals. However, non-organic farmers often manage their land well, encouraging species diversity and avoiding pollution. Below are listed some 'good practices' for farmers.

Planting hedgerows

This increases **species diversity** and some of the species attracted might well feed on potential pests, helping the farmer. Habitat variety can be further increased if the farmer leaves small areas of woodland on his land.

Preventing Soil Erosion

Top-soil blowing away reduces the fertility of the soil. Planting trees as **wind shields** can prevent this. This is particularly important in dry climates. Planting **hedgerows** also helps reduce soil erosion.

Using **organic fertilisers** can improve the soil structure, reduce pollution and help dispose of animal waste in a useful way.

Intercropping

This is the practice of growing two or more crops in the same field at the **same time**. It can produce a greater yield on a given piece of land, by using space that would otherwise be wasted with a single crop. Careful **planning** is required, taking into account the soil, climate, crops and varieties. An example is planting a deep-rooted crop with a shallow-rooted crop to make maximum use of soil nutrients. Intercropping encourages **biodiversity**, and it can also reduce pests — each crop may contain a chemical that repels the pest species of the other.

Planting Legumes

Legumes (peas, beans and clover) naturally **restore nitrates** to the soil. If planted alternately with other crops, there's less need to **fertilise** the soil. Peas and beans can be sold as a food crop, too.

Using Biological Control

This is an **alternative** to the use of chemical pesticides, and will be dealt with in the next section (see p. 74-75).

Reducing Eutrophication

Farmers can reduce the risk of eutrophication by adding fertiliser only when plants are growing and not to **bare soil**. It shouldn't be added when it is **raining**, and a strip of **unfertilised** soil should be left around the outside of the field. The soil should be **tested** before it is fertilised to check the nitrogen levels, and the amount of fertiliser carefully measured so that excess nitrogen is not put into the soil.

Practice Questions

Q1 Give four differences between intensive and extensive farming methods.

Q2 Why do intensive farming methods often result in the destruction of hedgerows?

Q3 What is the advantage of planting legumes in between two plantings of other crops?

Exam Questions

Q1 Describe the benefits and problems of intensive food production. [6 marks]

Q2 Explain the advantages and disadvantages of the use of organic fertilisers as an alternative to inorganic ones. [5 marks]

Q3 Explain what eutrophication is and how it may be caused as a result of farming practices. [8 marks]

Phew, that was a bit intensive...

Actually, that's not true. In fact, I'd say the page was pretty darn straightforward. Isn't legume a funny word though? You expect some kind of alien plant with legs... but no... they're just plain old peas and beans. Anyway, I feel we're losing track a bit, so what was I going to say... oh yes, the most important thing to make sure you understand on this topic is *Sorry — I've haven't got space to tell you.*

Diversity

This section is for Option Module 6 — Applied Ecology.
Diversity is all about how many different species there are in an ecosystem. In total, about 1.5 million species have been described but scientists reckon that the total number of species on Earth may be as high as 10 million.

*There is a **Link** between **Diversity** and **Stability** in an ecosystem*

Ecosystems with a **large diversity** of species tend to be **more stable** than those that are less diverse.
There are two ways to tell how **stable** an ecosystem is:

1. Stable ecosystems are **resistant** to **change**.
2. If **disrupted** in some way, stable ecosystems return to their **original state** quite quickly.

If you think about it it makes sense — **low diversity** means that **predator** species don't have much **choice** of prey.
If the population of a prey species is **reduced** or **wiped out**, then the predator species will be **at risk**.

When diversity is **higher**, the **predator** species will have a **large selection** of possible prey species.
If one of the prey species is wiped out, there will still be plenty of other species that predators can eat.

1) **Extreme** environments like **tundra, deserts, salt marshes** and **estuaries** are all ecosystems with **low diversities**. **Monocultures** are agricultural areas where only one crop is grown (e.g. wheat fields) — they have **artificially** low diversities. In areas of low diversity, plant and animal populations are mainly affected by **abiotic** (non-living) factors.
2) **Ecosystems** with **high diversities** are usually mature (i.e. old), and natural (e.g. oak woodlands), with environmental conditions that aren't too hostile. In these ecosystems populations are mostly affected by **biotic** (living) factors.

Biotic factors are things like **disease**, **predation** and **availability of food**. Biotic factors tend to be **density-dependent** — the bigger (denser) the population, the greater the effect the factor has on the population's size. For example, if a population has grown very big, a shortage of food will have more of an effect as there's more **competition** for it.

Density-independent factors have the same effect whether a population is big or small. They're often **abiotic** factors, like fire, flooding or temperature. For example, a spell of severe weather might reduce the size of a population whether it's near its carrying capacity or very small.

Diversity** is measured using a **'Diversity Index'

The simplest way to measure diversity is just to count up the number of species. But that takes no account of the **population size** of each species. Species that are in an ecosystem in very **small** numbers shouldn't be treated the same as those with **bigger** populations.

A **diversity index** is an equation for diversity that takes different population sizes into account. You can calculate the diversity index (**d**) of an ecosystem like this:

$$d = \frac{N(N-1)}{\sum n(n-1)}$$

Where...
N = **Total number** of organisms of **all** species
n = **Total number** of **one** species
Σ = **'Sum of'** (i.e. added together)

The **higher** the number the **more diverse** the area is. If all the individuals are of the same species (i.e. no diversity) the diversity index is 1.

Here's a simple example of the diversity index of a field:

There are 3 different species of flower in this field, a red species, a white and a blue.
There are 11 organisms altogether, so N = 11.
There are 3 of the red species, 5 of the white and 3 of the blue.
So the species diversity index of this field is:

$$d = \frac{11(11-1)}{3(3-1)+5(5-1)+3(3-1)} = \frac{110}{6+20+6} = 3.44$$

When calculating the bottom half of the equation you need to work out the n(n-1) bit for each different species then add them all together.

*A variety of **Microclimates** leads to **Higher Diversity***

Microclimates are **small areas** where the **abiotic** factors are **different** from the surrounding area. For example, the underneath of a rock has a different microclimate than the top surface — it's cooler and more humid.

Each microclimate provides a slightly different **habitat** that will suit **certain species**. So, ecosystems that have a variety of microclimates can support a **high diversity**. Basically: **more microclimates = more species = higher diversity**.

Diversity

For the Applied Ecology module, you need to know about frame and point quadrats and how to use them. So make sure you have a very good read of pages 52-53.

To sample **Animals**, you've got to **Catch** them

Remember, for your sample sites to be representative of the whole ecosystem, you must select them **randomly** (see page 54).

Most animals are **mobile** so they can't be sampled using quadrats or transects. There are various methods for catching animals depending on their **size** and the **kind of habitat** being investigated.

Nets can be used to trap flying insects and aquatic animals.
Pitfall traps can be used to catch walking insects on land. The insects fall into the trap and are... well, trapped.
Pooters are used to catch individual insects which are chosen by the user.
Tullgren funnels are used to extract small animals from soil samples. The animals move away from the light and heat produced by the bulb and eventually fall through the barrier into the alcohol below the funnel.

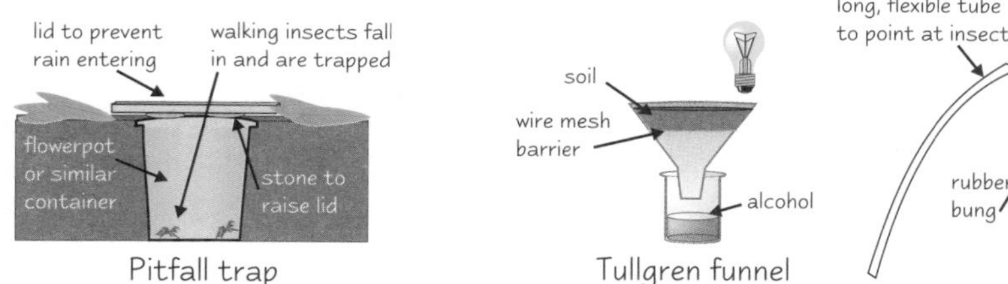

The **Mark-Release-Recapture** Technique is for estimating **Population Sizes**

The mark-release-recapture method is basically this:

1) **Capture** a sample of the population.
2) **Mark** them in a harmless way.
3) **Release** them back into their habitat.
4) Take a **second sample** from the population.
5) **Count** how many of the second sample are marked.
6) Estimate the **size** of the whole population using the **Lincoln index**.

$$\text{Population size (S)} = \frac{n_1 \times n_2}{n_m}$$

n_1 = number of individuals in first sample
n_2 = number of individuals in second sample
n_m = number of marked individuals in the second sample

The accuracy of this method depends upon these **assumptions**:

1) The marked sample has had enough **time** and **opportunity** to **mix** back with the population.
2) The marking has not affected the individuals' **chances of survival**.
3) **Changes** in population size due to **births**, **deaths** and **migration** are **small**.
4) The marking has **remained visible** in all cases — so it needs to be waterproof.

Good ways of marking animals include using a UV pen or cutting a little bit of the animal's fur off.

Practice Questions

Q1 What is meant by a 'stable' ecosystem?
Q2 Give three examples of ecosystems that are likely to be unstable.
Q3 Give an example of a diversity index.

Exam Question

Q1 A population of woodlice were sampled using pitfall traps. 80 individuals were caught. The sample was marked and released. Three days later, a second sample was taken and 100 individuals were captured. Of these, 10 had marks. Use the Lincoln index to estimate the size of the woodlouse population. [2 marks]

What do you collect in a poo-ter again?

Aren't you glad that we don't use the mark-release-recapture technique to measure our population size. I don't fancy falling in a pitfall trap and then getting a chunk of my hair cut off. Seems kind of barbaric, now I think about it. When we did this experiment at school, we never caught any of the woodlice we'd marked again. They'd all disappeared...

Pollution of Aquatic Ecosystems

Industry means manufacturing products or providing services in order to make a profit. Sometimes it can cause pollution, especially to aquatic environments — rivers, lakes and the sea. This has to be avoided wherever possible.

Heavy Metals cause Long-Term Problems in aquatic ecosystems

Heavy metals are those at the end of the periodic table (with high atomic masses). They're toxic, but due to a lack of alternatives many are still used in industry, and they sometimes get released into the environment in **industrial effluent**. The main heavy metal pollutants are **mercury, lead, cadmium** and **arsenic**.

1) **Mercury** is used in making **batteries**, and also in the **chemical industry**.
2) The use of **lead** as a **fuel additive** is now being phased out, but it still has widespread uses in the **construction industry** and in **electrical equipment**.
3) **Cadmium** and **arsenic** are also used in the manufacture of **electrical equipment**.

Heavy metals are toxic because they act as **enzyme inhibitors**. In smaller doses they can cause symptoms like **nausea** and **vomiting**, but in larger amounts they can be **fatal**. They're also **persistent** — which means that they don't break down in the environment or within the tissues of living organisms.

Bioaccumulation happens when they stay in an organism's living tissue. As the organism consumes more and more of the heavy metal, the levels in its tissues become **increasingly toxic**. They are passed from one **trophic level** to the next, and each time they become increasingly **concentrated**. The species that are highest up in the food chain (e.g. birds of prey) can end up with **lethal** concentrations in their bodies.

Mercury pollution in the **sea** is especially dangerous — microorganisms convert it into **methyl mercury**, which is even more toxic, and which accumulates in fish. Methyl mercury damages the **nervous system**.

Toxicity Tests can help Avoid the Release of Dangerous Substances

The **toxicity** of a substance can be measured in the laboratory, by finding either the **lethal dose** or the **lethal concentration**. This is done by giving groups of **test animals** (or plants) different amounts of the substance.

- The **lethal dose** (**LD50**) is the **amount** that **kills 50%** of the group tested (when injected or swallowed).
- The **lethal concentration** (**LC50**) is the **concentration** that **kills 50%** of the group (when breathed in).

This can help scientists to decide whether a new substance is safe to use. But it's obviously a test that should be used sparingly. As well as harming the test organisms, the results can be **misleading**. The lethal amount will depend on **how long** the test is carried on for, and on other conditions such as **temperature** or the presence of **other pollutants**.

Burning Fossil Fuels Causes Acid Rain

Aah, your old friend acid rain from GCSEs. Clearly it's a terrible problem, but isn't it lovely and reassuring to see something so familiar at A2...

Fossil fuels (coal, oil and natural gas) contain **nitrogen** and **sulphur** compounds as **impurities**. When these are burnt they produce **nitrogen oxides** and **sulphur dioxide**, which dissolve in rainwater and form weak nitric and sulphuric **acids**. This **acid rain** can be harmful to plants and animals:

- It **kills plants**, including trees, by damaging their **leaves** and by altering the pH of the **soil**.
- It **acidifies** lakes and rivers. The low pH kills animals and plants directly, but also causes **toxic aluminium ions** to be washed out of soil and into the water. On the other hand, **phosphate ions** (essential nutrients for plants) are **precipitated** as solid in acid conditions, and therefore sediment into the soil rather than remaining dissolved in the water.

Another side-effect of using oil as a fossil fuel is that **oil spillages** sometimes occur from tankers at sea. Oil damages wildlife simply by its **physical effect** — it sticks to seabirds and other organisms, **smothering** them and preventing them from **moving freely**. It can also **poison** them if they try to remove the oil from their bodies. The oil can also harm **shore** organisms, especially **shellfish**.

Pollution of Aquatic Ecosystems

Pollution Reduces **Species Diversity**

When water is badly polluted, most animals die because of the **low oxygen concentration**. Only a few **pollution-tolerant** species can survive the conditions, and these will thrive due to the lack of competition. This means polluted water has **fewer** species present — a lower **species diversity**.

Species diversity can be expressed as a **diversity index**, which uses a special formula to compare the number of different **species** in an ecosystem with the total number of **individuals**. There are **different versions** of this index — see page 62 for one common example.

Indicator Species can be Used to Check **Water Quality**

Some organisms can only survive in **clean water**, so if you found specimens of them in a sample of water it'd be **unlikely** that there was a problem with pollution there. Other species are adapted to cope with the extreme conditions found in **polluted water**, so if you saw a lot of them in a sample you'd know there was a problem. Organisms like this, that tend to be found in certain conditions, are known as **indicator species**. Examples include:

1) The **rat-tailed maggot**. This is found in water polluted with **organic matter**. It is adapted to survive in **low oxygen concentrations**.

2) ***E. coli***. This is a bacterium that is usually found in the human **large intestine**. Its presence in fresh water suggests pollution by **human sewage**.

3) **Freshwater shrimps**, **stonefly larvae**, **mayfly larvae** and **caddis fly larvae** are all indicators of **clean** water.

The presence of mermaids usually indicates clean water

Practice Questions

Q1 Name three toxic heavy metals which are still used in industry.

Q2 What does the term 'bioaccumulation' mean?

Q3 Explain the difference between a lethal dose and a lethal concentration of pollutants.

Q4 Describe the effect of acid rain on lakes and rivers.

Q5 Why does polluted water usually have a lower diversity index?

Exam Questions

Q1 A sample of water was found to contain rat-tailed maggots. What does this say about the water quality? [2 marks]

Q2 How might you know whether water is polluted with human sewage? [2 marks]

Q3 Name three species that are indicators of clean water. [3 marks]

Heavy metal can also cause long-term problems with your hearing...

That's a different type though, I think. Anyway, these pages should come as a welcome bit of light relief — good old acid rain, oil spillages, pollution and rat-tailed maggots (rats, maggots and organic waste all combined into one cuddly little friend). It's not too hard, and it's actually kind of interesting. I just love learning about organic pollution. Mmmm, slurry.

Adaptations to the Environment

Adapt to new conditions — or become extinct. Organisms that aren't well-suited to their environments can't survive. It's natural selection, and no species can avoid it.

The **Shape** and **Size** of Organisms are **Suited to their Environment**

Adaptations may be **structural**. For example, very hot (or cold) climates limit the **size** of organisms. In hot conditions, large organisms have difficulty losing enough **heat**, because of their low **surface area: volume ratio**. And in cold conditions, small organisms may lose **too much** heat.

But the **shape** of an animal will also affect its **surface area: volume ratio**. This is why elephants have such **large ears** — essentially, they have radiators on the sides of their heads, to improve heat loss. The large **surface area** of the ears means that the animal can **lose heat** more easily (it has special **blood vessels** to help with this).

By contrast, small mammals that live in cold conditions have **compact rounded bodies**, to minimise the surface area. Compare the body shapes of the Arctic hare, which lives in very cold conditions, and the brown hare, which lives in a more temperate climate. The more compact body shape of the Arctic hare and its smaller ears help it to retain heat more easily.

Some plants have **Structural Adaptations**

Xerophytes are plants that live in dry conditions. They're **adapted** to cope with the lack of water and avoid dehydration. Natural selection has meant that plants which are successful in arid environments are **efficient** at **water uptake** and have facilities to **reduce transpiration** and **store water**. Cacti are adapted to desert conditions:

- **Transpiration** is reduced by various leaf adaptations — a **thick waxy cuticle**, few stomata, sunken stomata, stomata that **open at night** and **close by day**, and a leaf surface covered with **fine hairs**.
- **Water uptake** is **increased** by having an **extensive root system** which covers a wide area.
- Water is **stored** in **fleshy**, **succulent** leaves or stems. Cacti also have **spines** to **protect** the plant from predators.

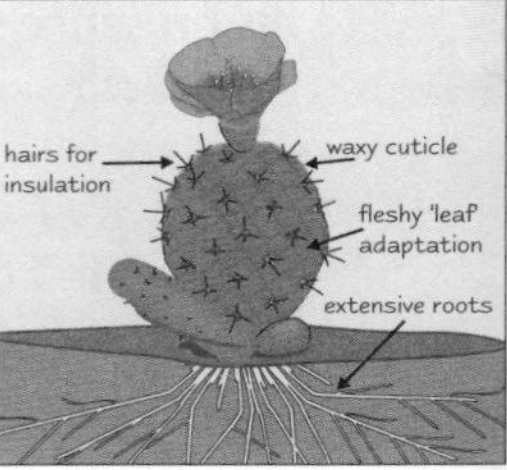

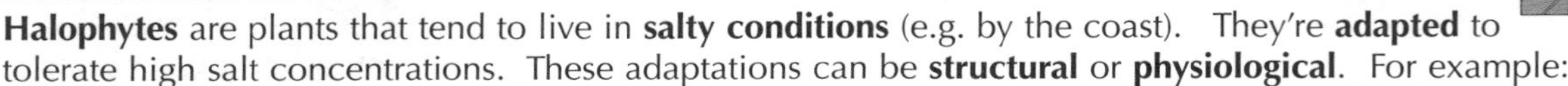

Halophytes are plants that tend to live in **salty conditions** (e.g. by the coast). They're **adapted** to tolerate high salt concentrations. These adaptations can be **structural** or **physiological**. For example:

- **Glands** that **secrete salt**, removing it from the plant. **Cordgrasses** (***Spartina***) have these.
- **Water stores**, so that the salt is **diluted** within the plant. Because of their water reserves, these plants look swollen and fleshy, so they're called **succulents**. ***Salicornia***, the **glasswort**, is an example.

Plants and animals also have **Physiological Adaptations**

As well as structural adaptations, animals and plants might also have **physiological adaptations**. This means that the body or its metabolism works in a special way.

Kangaroo rats live in the **deserts** of north America. They are so well adapted to the dry conditions that they never drink... pretty amazing, eh? The only water they ingest is a small amount of water from their food. Instead of drinking they use a combination of **physiological** and **behavioural** adaptations to **make** and **retain** water:

- Their **kidneys** have very long loops of Henle, so there's a greater **surface area** to accumulate **more sodium chloride** ions in the medulla (see page 16). This means their medullas have especially **low water potentials** so **more** water can be reabsorbed from the collecting ducts, making their **urine** more concentrated.
- They use the water produced in respiration — **metabolic water**. And they metabolise **fat** because it produces **twice** as much water as **carbohydrate** does.
- When they breathe, the water vapour in exhaled air **condenses** against the nasal passages, reducing water loss.

Tropical plants like **sugar cane** have a special type of photosynthesis called **C4 photosynthesis**. In C4 photosynthesis, **carbon dioxide** is fixed from the atmosphere to make a **4-carbon compound**, instead of the normal **3-carbon** compound of the **Calvin cycle** (see pages 10-11). This way of fixing CO_2 is much quicker and more efficient than the normal pathway, especially when CO_2 concentrations are **low**. This is an advantage because tropical vegetation is very dense, and so **competition** for CO_2 is intense.

Adaptations to the Environment

Animals can also have **Behavioural Adaptations**

Organisms may be adapted in their **behaviour** as well as in their structure and physiology. Any behaviour that improves survival chances is said to be **adaptive**. **Bird migration** is an interesting example. Migration lets birds exploit food resources in **two completely different areas**, depending on the season. Also, nearer the poles the summer days are **longer** (giving more time for feeding), but nearer the equator the winter days are both **longer and warmer**.

In Britain, many of our summer birds (e.g. **swallows** and **house martins**) migrate to **Africa** once the weather gets colder. In the summer, food is plentiful (and competition is less intense) in **Britain**. In winter, when the weather gets bad, the birds go **south** to warmer climates. Our winter birds, on the other hand, have migrated to Britain from even colder countries like **Scandinavia**. When these areas get covered with ice and snow, the migrants come south to Britain. Examples are **redwings** and **fieldfares**.

What triggers the migration behaviour of birds is a **combination** of factors, but all of these things may be involved:

- Birds have an **innate annual cycle** that means that they tend to move south (for example) after six months, and then move north six months later. This type of **genetic programming** is called an **endogenous cycle**.
- Migration is also affected by **day length**, which therefore **reinforces** the endogenous cycle. Birds migrate south when the days become shorter than a certain **critical period**. This kind of response to day length is called **photoperiodism**.
- In many species, migration can also be triggered by **reduction in food supplies**.

Once the migrating behaviour has been triggered, birds take in **extra food** to build up their fat reserves. The migration itself depends on navigation by **vision**, **wind direction** and **magnetic fields**. **Learning** and past experience of the route may also play a part, and younger birds may rely on more experienced members of the flock to guide them.

Simple Organisms show **Simple Adaptive Behaviours**

Even very simple organisms show **adaptive behaviour**.

1) **Tactic responses** (**taxes**) are responses where the organism moves towards or away from a **directional stimulus**. For example, **woodlice** show a tactic response to light (**phototaxis**). They move **away** from the light, which keeps them concealed under stones during the day (where they're safe from predators and in damp conditions).

2) **Kinetic responses** (**kineses**) are slightly different. Here, the animal's movement is affected by the **intensity** of the stimulus. Woodlice show a kinetic response to **humidity**. In **high humidity** they move **slowly** and **turn** more often (so that they stay where they are). As the air gets **drier**, they move faster and faster and turn less often, so that they move into a **new area**. The effect of this response is that the animals move from drier air to more humid air, and then stay put. This improves the **survival** chances of the animals — it reduces their water loss, and it helps to keep them concealed.

Practice Questions

Q1 Define the terms xerophyte and halophyte.

Q2 Name two physiological adaptations of the kangaroo rat which allow it to live in a desert environment.

Q3 Describe three different factors that might cause a bird to migrate south.

Exam Question

Q1 The smallest mammal in Britain is the pygmy shrew. It eats more than its own body weight in food every day. Larger shrews eat about half their own body weight each day. Suggest why:

a) The smaller shrews need to eat relatively more food. [3 marks]

b) A mammal smaller than the pygmy shrew could not survive in Britain, but small insects can. [2 marks]

You are the weakest organism... goodbye.

All these adaptations seem too clever to have happened just by chance, but remember that they evolve over millions of years. If a random trait appears due to mutation in a population and turns out to be an advantage, it'll then spread quickly. This is because if it helps an individual to survive, that individual will have time to reproduce more and pass the trait on.

Agricultural Ecosystems and Crop Production

Remember food chains? Bunny eats grass, fox eats bunny, blah blah blah. Plants are the producers in food chains. Their productivity is a measure of how much food they produce.

Productivity is a Measure of how much *Biomass* Producers make

Gross productivity (also called **gross primary production**) is the **total** amount of **energy** in the **biomass** made by the producers in a community. Some of this energy goes towards plant growth, but most gets broken down again in **respiration** to keep the plants alive. So we also have...

Gross productivity is often described as the amount of energy in the plant biomass **in a given area and in a given time**. It's given units like kJ m^{-2} $year^{-1}$.

Net productivity (**net primary production**). This is the amount of biomass that is left **after** respiration, which can be passed on to the next stage of the food chain. The equation for net productivity is:

Net productivity = gross productivity – respiratory loss

Leaf Area Index gives an Idea of *Plant Efficiency*

The **leaf area index** measures the **surface area** of the **leaves**, compared with the area of **ground** the plant covers:

Leaf area index = leaf area exposed to light ÷ area of soil surface

In measuring the area of a leaf, we only count **one side**. Even so, the leaf area index is usually **more than 1**. This is because leaves **overlap**.

Plants are most efficient when the leaf area index is about **4**. This gives enough overlap to absorb nearly **all** the light, but without **wasting** energy on leaves that are not needed.

Net Productivity is Less than *Gross Productivity*

Net productivity is the food that the farmer can actually **harvest**, so he or she wants this to be as high as possible.

Productivity depends on the **efficiency** of the plant — how good it is at changing **sunlight** energy into **chemical** energy.

For most crops, gross productivity is only about **1%** of the Sun's energy. So **99%** of the Sun's energy is **wasted**. Net productivity is even **less** — probably only about **0.1%**. Energy gets lost in all of the following ways:

1) Some of the Sun's rays may **miss the plant** and hit bare ground, or they may pass **straight through** the leaf.
2) Some rays are **reflected** (green light especially, that's why plants look green), but also **UV rays** and **infra-red**.
3) Some light may be absorbed by the **wrong bits** of the plant — e.g. by the cell walls instead of the chloroplasts.
4) The reactions of **photosynthesis** are not completely efficient, and energy is wasted as heat during photosynthesis.

All of these things reduce **gross productivity**. And **net** productivity is even lower because some food is broken down again by **respiration**, so the energy gets released as **heat**.

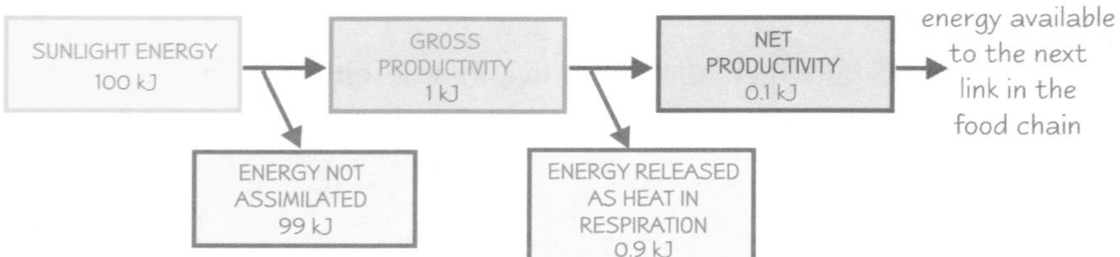

For every **100 kJ** they receive from the Sun, plants only pass on about **0.1 kJ** to the next stage of the food chain — that's **0.1% efficiency**. But efficiency **varies** according to the plant and the climate. Anything above **1% efficiency** is reckoned to be **pretty good**.

Modern **intensive farming** has led to a vast **increase** in the **net productivity** of both plants and animals. For example, in **glasshouses** the temperature, carbon dioxide concentration and light intensity can all be controlled to ensure that photosynthesis is as efficient as possible in the crops.

Intensive farming is **great** because there's never a shortage of food in this country. However, there are also lots of debates over intensive farming. For example, it has led a decrease in species and genetic **diversity**, which means that species might not have the **genetic variability** to cope with environmental changes in the future. There are also concerns over the welfare of farm animals, which are often kept warm and still to reduce their **energy use**. This means they don't need so much **energy input**, which saves money, but leads to more **diseases** from stress and overcrowding. Not good.

Agricultural Ecosystems and Crop Production

Fertilisers Increase **Productivity**, but **Too Much** is Money Down the Drain

Fertilisers make plants grow faster, because they give the plants essential minerals and nutrients. The most important mineral ions are **nitrate**, **phosphate** and **potassium**, but there are lots of **trace elements** needed in small amounts.

There are **two main types** of fertiliser. Each has its own advantages:

1) **Natural** fertilisers are **organic** matter (that's "muck" to you and me). They include manure and sewage sludge.
2) **Artificial** fertilisers are **inorganic**. They contain pure chemicals (e.g. ammonium nitrate) as powders or pellets.

Natural fertilisers supply a **wide range** of nutrients and release them slowly for a **long-lasting effect**. They're **less harmful** to the environment and are suitable for "**organic**" farming. They're also **cheaper** than artificial fertilisers and can improve the **soil structure**. However, they're expensive to **transport and apply**, and might not have the **ideal balance** of nutrients. They also contain **microorganisms** that use up some of the nutrients.

Artificial fertilisers are **fast-acting** and easy to transport and apply. They can be used to target **particular** mineral ion needs, and the amount of each mineral supplied can be **accurately controlled**. However, they're more **expensive**, can upset the **balance** of the soil, and are more easily washed out of the soil leading to **eutrophication**.

Farmers have to add the **right amount** of fertiliser to their crops, otherwise they're wasting money. Not enough, and the **yield** is reduced. But too much and the extra is just **wasted**. A **law of diminishing returns** comes into play — each extra amount of fertiliser makes less and less difference.

Crop Rotation Gets the Most from the **Soil**

Growing the **same crop** year after year in the same field can cause problems:

1) **Pests** and **diseases** start to build up. Their **eggs** or **spores** stay in the soil, ready to infect next year's crop.
2) If the crop needs a lot of one particular mineral ion, this will get **depleted**.

Farmers can get round this in one of two ways. They can either use **pesticides** and **fertilisers**. Or — the traditional method — they can use **crop rotation**. By **changing** the crop each year, crop rotation stops the pests and diseases of one crop building up in the soil. Also, it helps to stop mineral nutrients running out (because each crop has slightly **different** needs). Most crop rotations include a **legume plant** (like peas or beans), because these plants contain **nitrogen-fixing bacteria** in special **root nodules** in their root system, which means they increase the **nitrogen content** of the soil.

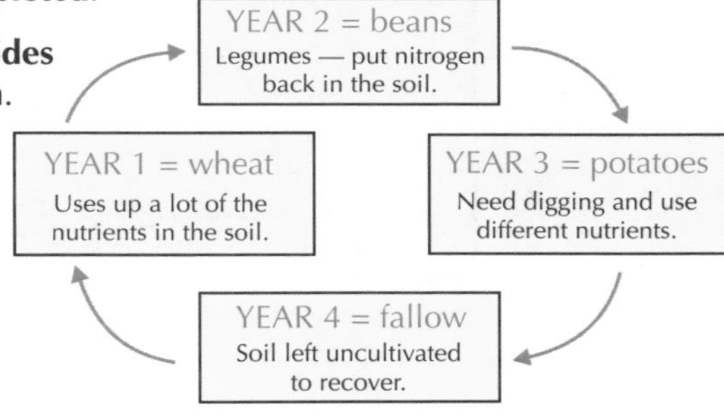

Nitrogen-fixing bacteria (e.g. ***Rhizobium***) use the enzyme **nitrogenase** to convert **nitrogen gas** and **carbohydrates** into **amino acids**. They get their sugars from the bean plant. In exchange, they give the plant **amino acids** for growth.

Practice Questions

Q1 What is the difference between gross productivity and net productivity?

Q2 Define the term leaf area index.

Q3 Give four reasons why only about 1% of the Sun's energy ends up as chemical energy in plants.

Q4 What is meant by the law of diminishing returns in fertiliser use?

Exam Questions

Q1 Gross productivity in a hot desert is about 750 kJ m^{-2} y^{-1}. In a tropical rainforest, it is about 80 000 kJ m^{-2} y^{-1}. Suggest three reasons for this difference. [3 marks]

Q2 Explain why net productivity is less if:
a) the leaf area index is less than about 3. [2 marks]
b) the leaf area index is more than about 6. [2 marks]

My brother's biology teacher thought his productivity was pretty gross...

She said it was disgusting how little work he did. Don't you be the same — instead get this page learned, as efficiently as possible. You'll probably need to stop reflecting rays and absorbing light with your cell walls. Those photosynthesis reactions are wasting heat, and as for your leaf area index... no, it's no good — I'm going to have to cover you with manure, I'm afraid.

Harvesting Ecosystems

Fishing isn't like farming. Farms are artificial, designed just for food production. But fishing is taking food from a natural ecosystem. If the ecosystem isn't carefully conserved, the food could run out.

Fisheries Aim for Maximum Sustainable Yield

Fish are able to reproduce very **rapidly** if the conditions are right. For example, a single pair of **cod** can produce **four million** fertilised eggs every year. But the young fish need time to develop, and their **mortality rate** is high even from natural causes. If fishermen take too many fish, it could cause the population to **collapse**.

Overfishing is where the fish can't reproduce fast enough to replace what the fishermen remove. For one or two years, the fishermen may get an excellent yield from the sea, but this yield will then drop dramatically — it's **not sustainable**.

Underfishing is where fishermen take fewer fish than they could. This is OK for the fish, of course — fewer of them die — but it means that a potential food resource is going to **waste**.

The **maximum sustainable yield** is the largest amount of fish that can be caught without causing the population to fall. It's obtained when the fish population is conserved, and the **population growth curve** is kept at its **steepest** part. In this way, the population is made to keep producing food efficiently. With **underfishing**, there are lots of large, older fish which aren't growing much, and which eat food that could be used for growth by younger fish. With **overfishing**, few individuals reach maturity, so the **breeding stock** decreases and fewer young fish are produced.

A population growth curve looks like this:

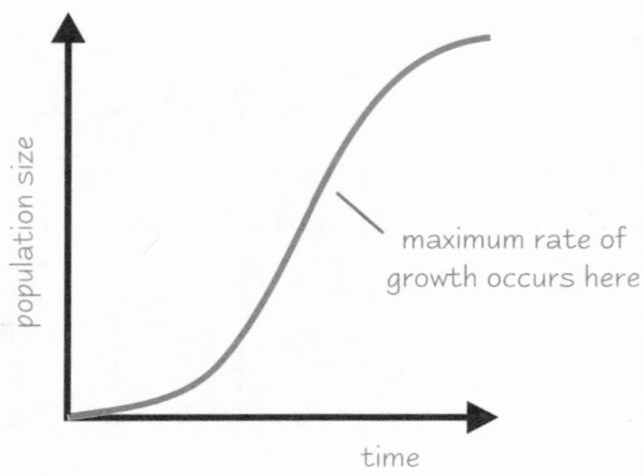

The maximum sustainable yield is obtained by keeping the population at the level where maximum growth occurs:

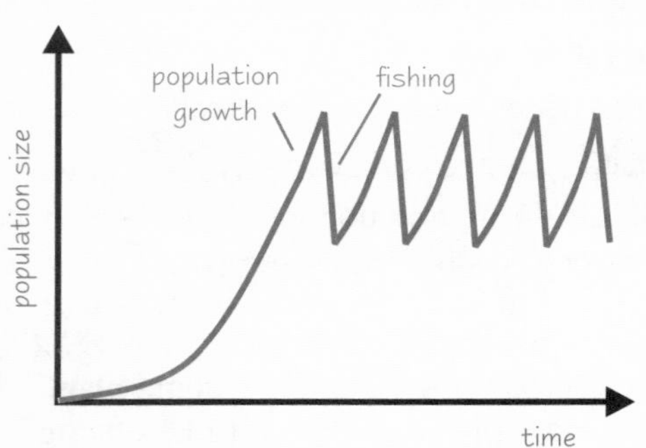

Modern Fishermen are Too Good at their Job

Modern fishing methods are extremely **effective**. So effective that if all the fishermen took as many fish as they could, the fish population would soon be **exterminated**. So, in popular fishing areas like the North Sea, **regulations** have been introduced to prevent **over-fishing**:

1) **Quotas** — each boat is limited to a **maximum weight** of fish that it's permitted to catch.
2) **Exclusion zones** — parts of the sea are made **off limits** for fishing boats, so that there are areas where the fish can breed safely.
3) **Close seasons** — fishing may be forbidden at certain **times of year**, usually the **spawning season**. This gives the population a chance to **recover** through reproduction.
4) **Net size restrictions** — fishing with excessively **large** nets may be forbidden. Also, fishing boats may be required to use nets with a **larger mesh size** (e.g. 90 mm minimum for North Sea cod fisheries). This prevents them catching the **smaller** fish, so that young fish have a chance to grow and **breed** before they're caught.

All these methods help to conserve fish stocks, but there are sometimes problems **enforcing** the rules. For example, if fishermen have **quotas** they might still catch more fish than they're allowed to, then just keep the best (biggest) fish, and throw the smaller ones (already dead) back into the water. Fishing restrictions are also **politically unpopular**, and depend on **international co-operation**.

Harvesting Ecosystems

Fish Farming Can Increase the Harvest

In a **fish farm**, fish are raised in a controlled situation. This helps to **maximise the yield**. Fish farms may be **open** systems or **closed** systems:

- **Closed systems** are large tanks in which the water is continuously recirculated. This gives a lot of **control** over the conditions, but it's **expensive**. It might not be suitable for **marine** fish, especially **carnivores** that swim over a **wide area**.
- An **open system** is a section of the open sea (or river or lake) where **nets** restrict the movement of fish, but allow water to flow freely in and out. This is **cheaper**, but gives **less control** over the conditions.

The **yield** is increased under the **controlled conditions** in a fish farm because:

1) The fish are **protected** from **predators** like seals and carnivorous birds.
2) The fish are given a controlled **diet**, often in pellet form. Food is supplied in the right amount and at the right time to **maximise growth**.
3) **Competitors** are excluded, so the fish get all the food for themselves.
4) Special tanks can be used for raising the **young** in **ideal conditions**.
5) **Diseases** are controlled by the use of **pesticides** or **antibiotics**.
6) If the fish farm uses a **closed system**, the **temperature** and **composition** of the water can also be controlled.

Fish Farms Can Affect the Marine Ecosystem

Fish farming has some **disadvantages**. Some consumers think that the **quality** of the fish isn't as good, and fish farming is also relatively **expensive**. Fish farms can also damage the **environment**:

1) They produce a lot of **food** and **faeces**, attracting **scavengers** and **bacteria**. These then **deoxygenate** the water.
2) Fish farms can create high concentrations of **pests** and **diseases**. These pest species may then go out into the **open sea** and affect the **wild population**.

Practice Questions

Q1 Define the terms overfishing and underfishing.

Q2 Explain the meaning of the term maximum sustainable yield in relation to a fishery.

Q3 Give four ways that overfishing can be prevented.

Q4 Give three advantages and three disadvantages of producing salmon in a fish farm, rather than catching them wild.

Exam Questions

Q1 In the North Sea, fishermen are making use of a natural ecosystem. What are the advantages and disadvantages of this, compared with getting food from a farm on land? [4 marks]

Q2 In fish farms, bacterial diseases are often controlled using antibiotics. Suggest what damaging effects these antibiotics might have on the environment. [2 marks]

Fishermen are just too ef*FISH*ent... geddit? geddit?

Sorry. It's a touchy subject this — lots of fishermen are pretty cross about all these restrictions, and they say the scientists are talking rubbish and there's plenty of cod left. But how weird would it be if one day, the humble cod was to be found only in the zoo. And you'll take your grandkids on a day out, and you'll say "Look kids — I used to eat them with chips."

Conservation of Species

We don't just conserve species to be nice. And it's not because they're cute or fluffy. There are lots of very serious reasons why we should maintain biodiversity, which you need to know all about.

Biological Conservation Maintains Biodiversity

It's not because they're cute or fluffy. Not at all.

Conserving other species seems the right thing to do. But these other organisms are also vital as part of the **ecosystem** (extinction of one species affects everything else). And they're a **resource** — living things are used by humans as sources of **food**, in **medicines**, and as tools for **biotechnology**. Once an organism is extinct, its genes are eliminated forever from the gene pool, and are no longer available to us. This is particularly significant now, because **DNA technology** means that **any** gene has potential for useful exploitation.

You can **classify** conservation efforts in different ways. Here are the five you need to know about:

1) **Biological** conservation — maintaining the diversity of living organisms **within habitats**.
2) **Environmental** conservation — conserving the **abiotic** (non-living) characteristics of ecosystems.
3) **Nature** conservation — the preservation of sites of **special scientific interest**.
4) **Species** conservation — the protection of **rare species**.
5) **Global** conservation — conserving the composition of the **atmosphere** and **oceans**.

Conserving Species Means Conserving Habitats

Some species are under threat because their natural **habitats** are shrinking. This happens for **two** main reasons:

- Human populations **expand**, so habitats are cleared to make way for homes etc.
- Over-exploitation of **resources**, e.g. through deforestation or mining.

Habitat conservation is important for its own sake, and also for the sake of the organisms that live there.
By maintaining the **abiotic** characteristics of an ecosystem, the **biotic** component (the living organisms) can thrive.
In Britain, several measures are in place to promote habitat conservation:

1) **National Parks** — There are **fifteen** National Parks in Britain, ranging from Dartmoor to the Cairngorms. These regions are controlled by **Acts of Parliament** to conserve their natural beauty and to protect the organisms that inhabit them, but also to promote opportunities for the public to enjoy them.

2) **Sites of Special Scientific Interest (SSSIs)** — These are smaller areas (there are over **5000** SSSIs in Britain), often **privately owned**, which are particularly important because of their wildlife or geology. SSSIs are chosen by **English Nature**, by the **Countryside Council for Wales**, or by **Scottish National Heritage**. Once designated, there are **strict controls** on any proposed developments within them.

3) **Environmentally Sensitive Areas (ESAs)** — These areas are **nationally significant** places of ecological interest, where modern farming practices are threatening the environment. In ESAs, the government provides **financial incentives** to farmers to adopt environmentally friendly land management practices.

4) **Wildlife reserves** — Organisations such as the **Royal Society for the Protection of Birds** maintain and protect areas for **rare species** to live and breed safely.

Species can also be Conserved Outside their Natural Habitats

Even when habitats are threatened, some measures can be taken to conserve species:

- **Zoos** run captive **breeding programmes** for endangered species of animals.
  After successful breeding, the new generation can be **released** into the wild.
- **Botanical gardens** maintain **seed banks** — stores of the seed of threatened plants. They can be **germinated** later, and with careful storage they can be kept for an almost **unlimited** time.

Conservation of Species

A Case Study — African Elephants

There are only about **500 000** elephants in the whole of **Africa**. This may seem a lot, but it's down from about **4 million** in **1900**. If the numbers fall too far, **inbreeding** may weaken the population further.

Elephants are not just attractive animals, they're also an important part of the **forest/ grassland ecosystem**. If their population crashes, this will affect many **other** species too.

Why has the elephant population fallen?

- The growing **human** population in Africa has led to elephant **habitats** being cleared.
- The **ivory trade** has meant elephants are **hunted** for their tusks.

So far, conservation has concentrated on limiting the **ivory trade**. Since 1990 this trade has been **illegal**, and poaching is now more **strictly policed**. Measures like this need **international co-operation**, and the ivory trade is included in the **Convention on International Trade in Endangered Species** (**CITES**).

Tropical Rainforests are Important for the Whole Biosphere

Tropical rainforests are important for **two** main reasons:

1) Their **biodiversity**, including many species found **nowhere else** on Earth.
2) Their effect on **global climate**. The rainforests absorb **carbon dioxide**, and therefore help prevent global warming.

In the past hundred years, over **50%** of the world's tropical rainforests have been destroyed, and this destruction still continues. **Why** has it happened?

- Clearing the forest for **agriculture**, **industry** or **living space**.
- Logging of trees for their **valuable wood**. Something to think about next time you sit on a mahogany loo seat!

Trees, of course, re-grow in time, so some logging activities are **sustainable**. But if over-exploited, the rainforest destruction becomes **irreversible**.

Conservation measures for rainforests include:

1) **Educating** local people about how to manage the land for a sustainable timber yield.
2) **International agreements** on managing forests, including a scheme to **certify** timber from well managed forests.
3) Developing **alternatives** to tropical woods — softwoods from managed woodlands elsewhere, or completely different materials like plastics.
4) Laying aside **conservation areas** as completely free from logging.

As with many conservation measures, controls need co-operation from **local people** and **international agreement**.

Practice Questions

Q1 Give one example of a conservation measure for each of these types of conservation:
a) habitat conservation b) nature conservation c) species conservation

Q2 Explain the difference between a SSSI and an ESA.

Q3 Give two reasons why the conservation of tropical rainforests is important.

Exam Questions

Q1 Explain briefly why the environment has come more under threat over the past 100 years than in the previous 1000 years. [2 marks]

Q2 National parks are set up to protect wildlife, and also to provide opportunities for the public to enjoy them. Suggest how these two objectives might be in conflict with each other. [4 marks]

Q: How did a Chinese breeding programme help a giant panda get pregnant?

A: they showed her sex education videos. Seriously. Staff at the Wolong Giant Panda Protection Centre in south west China realised that the captive-born 4 year-old female, Hua Mei, had no knowledge of normal panda mating behaviour. So they showed her videos of mating pandas before a series of "blind dates" with males. What's even weirder is that it worked.

Controlling Pests

We had a pest in our cupboard once. He was a mouse and he ate all my rice so we called him Paddy.

There are **Three** methods of **Pest Control**

Pests are species that negatively affect human activities. They can be **weeds** that out-compete crops, **insects** that carry diseases or **fungi** that reduce crop yields. They could also be animals such as **rabbits** or **deer** that eat crops. There are three main approaches to dealing with pests:

1) **Chemical control** uses chemicals like pesticides to kill the pest species.
2) **Biological control** uses living organisms to control the pest. These include **predators, parasites** and **pathogens** which kill the pest or prevent it from reproducing.
3) **Integrated pest management** mixes both chemical and biological control together to combat pests.

Chemical Pesticides are pretty **Effective**

1) **Herbicides** are used to combat weeds. There are two kinds:
    - **Contact herbicides** kill weeds when they are sprayed onto their **surface**.
    - **Systemic herbicides** have to be **absorbed** into the weed to kill it. High doses of synthetic **auxins** (plant hormones) are used as systemic herbicides. Auxins are **selective** weedkillers, which means they kill weeds without harming crops — clever.
2) **Fungicides** are used to fight fungal infections. Like herbicides they can be either contact or systemic.
3) **Insecticides** are used against... guess what... insect pests. There are three types this time:
    - **Contact insecticides** have to come into direct contact with the insect.
    - **Systemic insecticides** are absorbed by the plant and carried in its **phloem**. They kill the insects that feed on the plant's sap.
    - **Stomach ingestion insecticides** are sprayed over the crop and are **consumed** when pests eat the plants.

Chemical Pesticides can cause **Problems**

Some pesticides are **persistent** (non-biodegradable) — which means that they don't break down in the environment or within the tissues of living organisms. This causes big problems for species that aren't supposed to be damaged by the pesticide:

Bioaccumulation happens when persistent pesticides stay in an organism's living tissue. As the organism consumes more and more of the pesticide the levels in its tissues become **increasingly toxic**. The pesticides are passed from one **trophic level** to the next, and each time they become increasingly **concentrated**. The species that are highest up in the food chain (e.g. birds of prey) can end up with **lethal** concentrations in their bodies.

Pesticides that aren't <u>specific</u> cause lots of problems because they kill other species, as well as the ones that they're supposed to attack.

Biological Control can be Better for the **Environment**

Biological control avoids the need for putting extra **chemicals** into the environment. Organisms that are used in biological control include:

1) **Insect parasites** which are specific to the pest. Most lay their **eggs** on or in the host and then when the eggs hatch the larvae eat the host from the **inside**. Nice...
2) **Predators** are **carnivorous** species used to control insects and other animal pests.
3) **Pathogens** are **bacteria** and **viruses** used to kill pests. For example, the bacterium ***Bacillus thuringiensis*** produces a **toxin** which kills a wide range of **caterpillars**.

Controlling Pests

Biological control has **Advantages** and **Disadvantages**

Advantages

1) The control organism is usually **specific** and only the pest species will be affected.
2) **No chemicals** are used, so problems of bioaccumulation and pollution are avoided.
3) Control organisms usually establish a population so there is **no need** for **re-application**.
4) Pests don't usually develop genetic resistance to the biological control agent (this is a big problem with chemical fertilisers).

Disadvantages

1) Control is **slower** than using chemical pesticides because you have to wait for the biological control agent to establish a large enough **population** to control the pests.
2) Generally biological control **doesn't permanently exterminate** the pest — instead it's reduced it to a level where it is no longer a big problem.
3) It can be **unpredictable** — it's really hard to work out what all the **knock-on effects** of the introduction of the biological control species will be.
4) A lot of **scientific research** into the relationship between the pest species and the biological control agent is needed. This research takes **time** and costs **money**.
5) If the control species has several sources of food its population levels might **grow** and it may **become a pest** itself.

There are **Other Ways** of dealing with **Pests**

In recent years farmers have started to use **integrated pest management** where chemical and biological methods are combined.

Biological control is used to keep the pest down so that it doesn't affect the **profitability** of the farm. If a pest **outbreak** occurs then the farmer uses a specific pesticide for a **short** period of time.

Crop rotation can help control pests — changing the crop each year makes it difficult for pests that can only feed on one species to become established.

Genetic modification can be used to produce crops which are **resistant** to pests, so no pest control is required. (Although it is possible for pests to evolve so they can feed on the genetically modified organisms.)

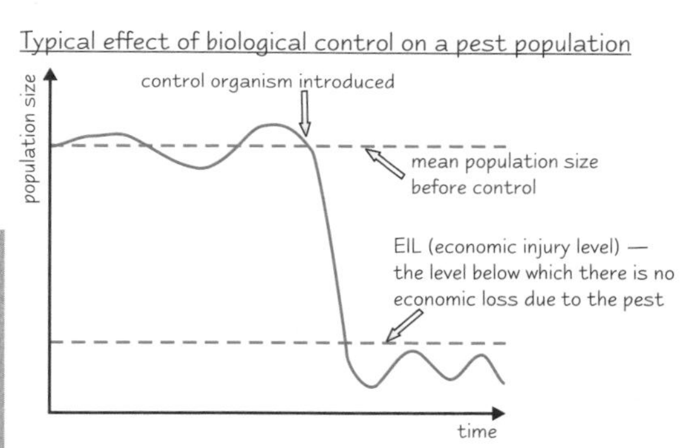

Practice Questions

Q1 What is the difference between a contact pesticide and a systemic pesticide?

Q2 What problem is caused in food chains by persistent pesticides?

Q3 What is integrated pest management?

Exam Questions

Q1 A chemical company is developing a new weed killer. State three features that the company should aim to include in their herbicide to ensure that it causes as little environmental damage as possible. [3 marks]

Q2 Explain why some farmers prefer to use pesticides rather than biological control to deal with pests. [5 marks]

There's no need to be pestimistic, this stuff's easy...

When you realise how difficult it is to control pests you can understand why some people think that genetic modification is the way forward. The thing that worries people is that, like biological control, it's hard for scientists to know exactly what the consequences of growing GM foods will be in the long term. It could be a great thing, or it could be a scary mistake.

Bacteria

This section is for Option Module 7 — Microbes and Disease.
*A **microorganism** is basically any organism that can only be seen with a **microscope**.*
They're mostly the single-celled type of beastie — bacteria, fungi, protoctista and the like.

Bacteria are Prokaryotic Microorganisms

All bacteria are **single-celled**, although some kinds of bacteria can stick together to make **clusters** or **chains**. Bacteria are really, really **small** — they range from about 0.1 to 2μm.

Here are the **features** shared by **most** bacterial cells:

1) They have a **cell wall** containing **peptidoglycan**.

2) They divide by **binary fission** (see below).

3) They **don't** have **membrane-bound organelles** (e.g. a nucleus), so **can't** do endocytosis or meiosis.

4) Their **DNA** consists of a circular molecule with **vital** genes (the bacterial **'chromosome'**), and smaller rings called **plasmids**, which contain other, non-vital genes, e.g. for antibiotic-resistance.

5) Since there's no nucleus, **transcription** occurs in the **cytoplasm**. **Translation** occurs on **70S ribosomes** (smaller than the 80S ones in eukaryotes).

6) **Genetic recombination** can occur by **conjugation**, where two cells join up to exchange parts of their DNA. Genes can also be **recombined** by exchanging plasmids, by transfer in viruses called **bacteriophages**, or by taking up DNA from the **surroundings** (e.g. from dead bacteria).

features in most bacteria

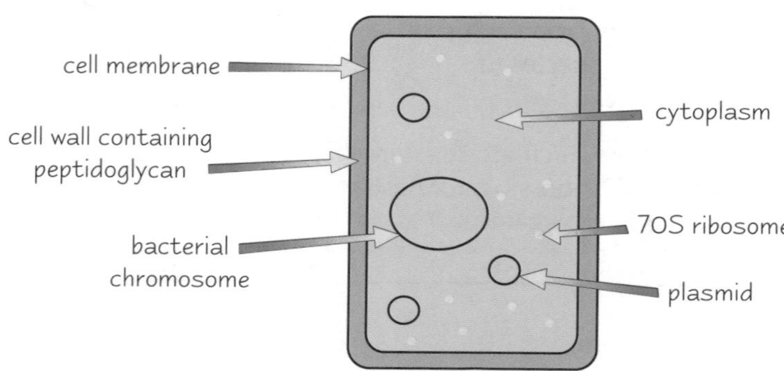

Bacteria divide by Binary Fission

Bacteria reproduce using a process called **binary fission**.
This basically means the bacterium splits in two to produce **clones**.

The DNA in the parent cell undergoes **replication** before it splits in two.

Binary Fission

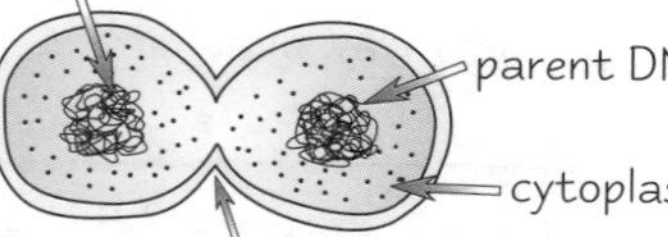

Bacteria

Some bacteria have **Special Features**

Here are some **special** features that are found in **only some** kinds of bacteria:

1) **Bacterial flagella** may be present.

2) **Mesosomes** are **folds** of the cell membrane — their purpose isn't known for sure, but they may be associated with respiration.

3) **Bristle**-like projections called **pili** and **fimbriae** allow the bacteria to **stick** to surfaces. (Fimbriae are **shorter** than pili and there are **more** of them.)

4) There may be a layer of **polysaccharide** (called a **glycocalyx**) around the cell wall. It could be a loose **slime** layer or a more solid **capsule**. It could **defend** the bacteria from white blood cells or stop it **drying out**.

5) Bacteria can be **heterotrophic** (consume organic molecules), **photoautotrophic** (make food using light energy) or **chemoautotrophic** (make food using energy from oxidation reactions).

6) Bacteria can be classified as either **Gram-positive** or **Gram-negative**. Gram-positive bacteria have a **thick** cell wall. Gram-negative bacteria have a **thinner** cell wall coated with an outer **membrane**.

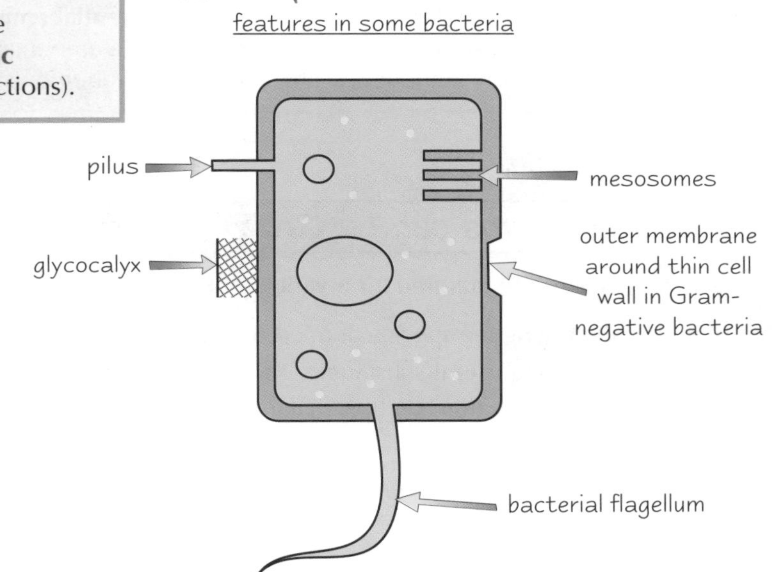

features in some bacteria

Practice Questions

Q1 Roughly how big are bacteria?

Q2 List three features shared by most bacteria.

Q3 What is conjugation?

Q4 What is the purpose of pili and fimbriae?

Q5 What is a glycocalyx?

Q6 List three ways in which bacteria might feed.

Exam Questions

Q1 Explain how genetic recombination occurs in bacteria. [4 marks]

Q2 Describe two possible functions of the layer of polysaccharide around a bacterial cell wall. [2 marks]

Bacteria — Unhygenix's wife...

A lovely couple of pages on bacteria. Mmmm... bacteria. A lot of this stuff shouldn't be too new. You'll know a lot about bacteria from all your past Biology studies. It makes it a bit easier for you, but it doesn't mean you can get by with what you know already. So go on — learn those nasty, long words like 'chemoautotrophic'. You know you want to.

Culturing Microorganisms

Loads of kinds of microorganism can be grown in the lab in cultures. Which means you have to learn all about it.

Microorganisms have Special **Chemical Requirements**

Autotrophic organisms (plants, algae and some kinds of bacteria) need a supply of simple inorganic molecules with which to build organic molecules. So, if they're being grown in a lab, there should be a ready supply of these things. Here are the **elements** they need and where they get them from:

1) **Carbon** and **oxygen** (in all organic molecules) are supplied by **carbon dioxide**.
2) **Hydrogen** (also in all organic molecules) is supplied by **water**.
3) **Nitrogen** (in amino acids and the bases of nucleotides) is supplied by **nitrate**, or sometimes by ammonia or nitrogen gas.
4) **Phosphorus** (in nucleotides) is supplied by **phosphate**.
5) **Sulphur** (in just two kinds of amino acids) is supplied by **sulphate**.

Heterotrophic organisms (animals, fungi and most bacteria) take in organic molecules, and some inorganic materials. They can make a variety of other types of organic molecules through their metabolism, depending on which enzymes they have.

Chemicals that microorganisms need in order to grow are called **essential nutrients**. For heterotrophs, these include **essential amino acids** and **essential fatty acids**. **Macronutrients** are needed in large amounts, but **micronutrients** are needed in tiny amounts. Many essential micronutrients needed by heterotrophs are **vitamins**.

Microorganisms can be **Grown** on **Different Media**

Microorganisms are cultured for a variety of **reasons**, including:

- **identifying** the species of microorganism in a medical sample — to **diagnose** the disease
- producing **useful** substances, such as **antibiotics**
- **food production**, such as making cheese and wine

In the lab, microorganisms are cultured **in vitro** ('in glass', like on an agar plate). The material where microorganisms are grown is known as the **medium**. It can be liquid, in the form of a **nutrient broth**, or solid, as in **agar jelly**.

Agar jelly is made by heating a solution of agar, a polysaccharide which solidifies to jelly when cooled to 42°C. Special nutrients are added before the agar sets.

Selective media contain substances which selectively **prevent** the growth of certain microorganisms, whilst allowing others to grow. This means that specific types of microorganism can be **isolated** and **cultured**.

Indicator media contain a chemical **indicator** (e.g. pH indicator) that changes colour due to chemicals made by certain types of microorganisms. For example, **EMB agar** contains lactose, sucrose, eosin and methylene blue. Methylene blue inhibits Gram-negative gut bacteria, and only some types of Gram-positive ones can use lactose or sucrose. Of these, *Escherichia coli* produce black colonies with eosin, whereas *Salmonella* produce pink ones.

Disinfectants and **Antibiotics** Affect **Growth** of Bacterial Lawns

The **effectiveness** of antibiotics and disinfectants at preventing bacterial growth can be discovered by adding them to bacterial lawns on sterile discs of filter paper.

The disinfectant or antibiotic **diffuses** out of the disc to create an **inhibition zone**, where bacteria can't grow. The **bigger** this is, the more **effective** the antibiotic or disinfectant.

Culturing Microorganisms

Aseptic Technique is used to Culture Microorganisms

Aseptic technique is handling a microorganism culture in a way that minimises **contamination**. **Sterilisation** is the removal or killing of **all** microorganisms on an object, so that after sterilisation, there aren't any microorganisms to reproduce, even if the most favourable conditions occurred. The equipment and media used are sterilised by heating to 121°C for 15 minutes in an **autoclave**.

Other precautions taken to minimise **contamination** by other microorganisms include:

1) Lab workers make sure they have **clean hands**, and wear a **lab coat**, **gloves** and sometimes a protective **face mask**.
2) Windows and doors are kept **closed**.
3) All containers holding microorganisms have **lids** to minimise contamination by **'fall out'** of microorganisms from the **air**.
4) Working close to a **Bunsen flame** ensures that many of the microorganisms in 'fall out' are **killed** before they land on working surfaces.
5) Surfaces are swabbed with **disinfectant**.
6) Culture **containers** are made of smooth glass or stainless steel as **scratches** can harbour stray microorganisms.
7) The microorganisms can be transferred from medium to medium on **wire inoculating loops** that are first **sterilised** by heating them to red hot in a Bunsen flame. Alternatively, a sterilised **Pasteur pipette** can be used for transferring a known volume of culture (necessary for quantitative work), and the culture is then spread on agar using a sterilised **spreader** made from a bent glass rod. This gives a fairly even **lawn** of microorganisms.

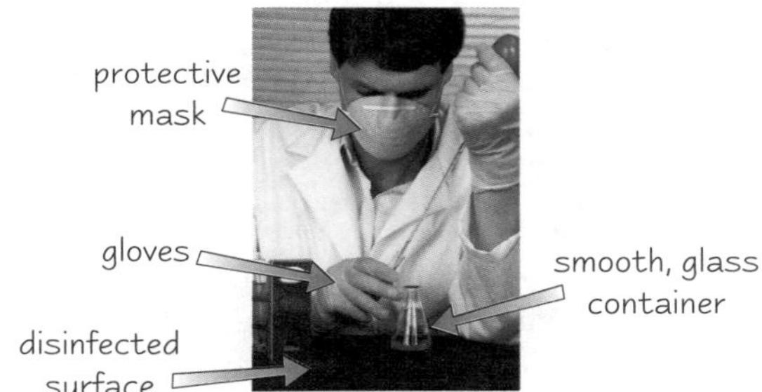

Practice Questions

Q1 Name five elements which microorganisms need to build organic molecules. Where do they get them from?

Q2 Give three reasons why microorganisms are cultured.

Q3 Explain the meaning of the terms 'selective media' and 'indicator media'.

Q4 How can you test how good a disinfectant or antibiotic is?

Q5 What is meant by sterilisation of laboratory apparatus?

Exam Questions

Q1 A student was provided with a monoculture of the mould *Aspergillus nidulans* on an agar plate in a Petri dish. She was asked to transfer a sample of the *A. nidulans* culture onto a separate sterile agar plate.

a) Describe the stages in the aseptic transfer of the sample from the monoculture to the sterile new plate. [5 marks]

b) Suggest a method that was used to sterilise the agar plate. [2 marks]

Glasgow, City of Culture 1998 — Great conditions for bacterial growth...

This is the sort of stuff that's really boring... but really important. I know you want to get on with learning about nasty diseases and other gruesome stuff, but you've got to take the time for cell cultures. If it wasn't for cell cultures, we wouldn't be able to diagnose lots of diseases, or make antibiotics to cure them, or, erm, make cheese. So there.

Bacterial Growth

It's a lot easier to measure the population growth rate of a bacterial culture, than of, say, lions. Or whales. Or people.

Bacterial Cultures Show Typical **Population Growth Curves**

Following the initial introduction of bacteria into the medium, the following **phases** of **population growth** are seen:

1) The **lag phase** occurs when there is a very **small** early **increase** in population density. **Reproduction rate** is **low** because it takes time for enzymes to be made for DNA replication and to use any new food.
2) The **exponential phase** occurs during the most **favourable** conditions. Here there is a **doubling** in population size per unit time (one cell divides to produce two cells, each of which can divide further in the same amount of time). There is sufficient **food** to support the growth, and **competition** for food is at a minimum.
3) The **stationary phase** is approached as **death rate** increases and becomes **equal** to the reproductive rate. This happens because **food** gets used up and poisonous **waste products** build up.
4) **Decline phase** occurs where death rate is **greater** than reproductive rate, due to the further depletion of food and accumulation of excreted waste.

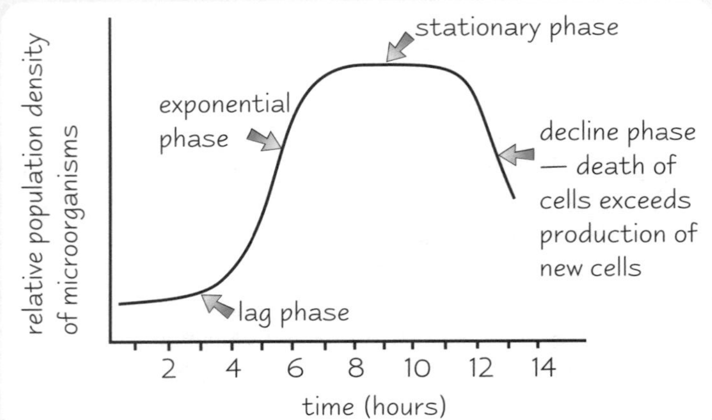

Depending on culture conditions, many bacteria produce substances called **secondary metabolites** towards the end of the exponential phase and into the stationary phase. These are not essential for growth and reproduction, but some, such as antibiotics, can help bacteria **survive** the stressful conditions at this point in the population growth curve.

Special **Conditions** are Needed for **Optimum Growth**

1) The optimum **temperature** for growth varies between microorganisms, with some having extremes. **Psychrophiles** grow best at low temperatures, **thermophiles** grow best at high temperatures.
2) Some microorganisms are **aerobic**, but others are **anaerobic**. **Obligate aerobes** can only respire aerobically and so will die if oxygen is not present. **Obligate anaerobes**, however, die in the **presence** of oxygen. If necessary, oxygen concentration in a culture can be increased by **aeration**.
3) Most species have an optimum **pH** of between 5 and 9, but **acidophilic** forms have a low optimum pH. Usually fungi are more tolerant of low pH than bacteria.
4) **Disinfectants** inhibit growth of microorganisms, and **antibiotics** kill them or inhibit their replication.

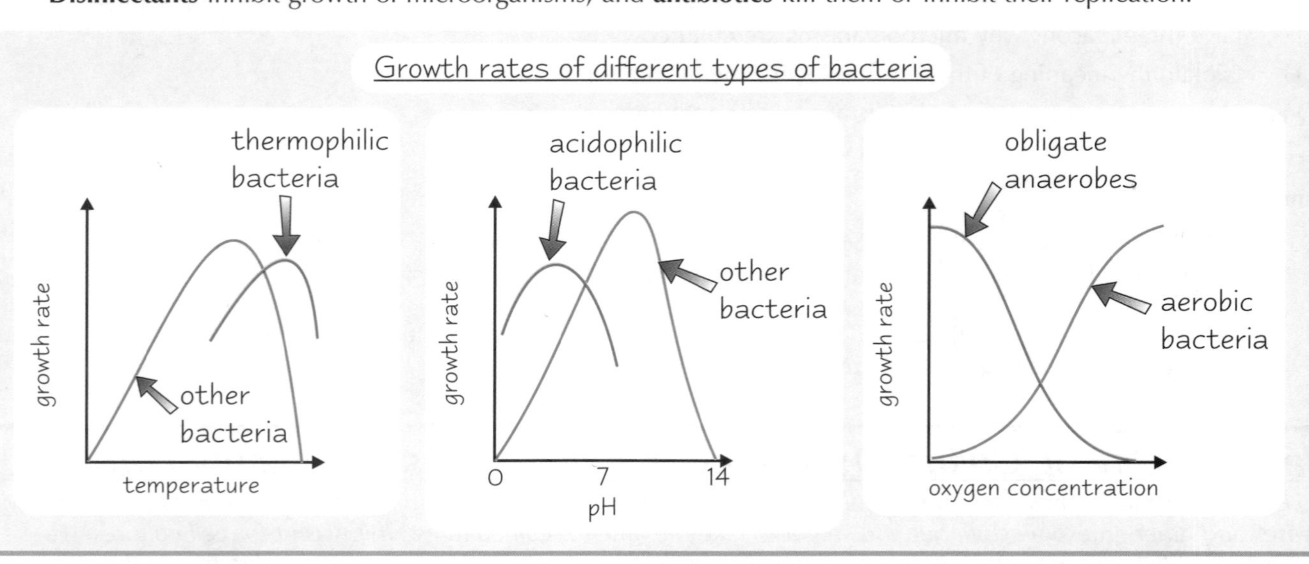

Bacterial Growth

There are Different Ways to Find out **Population Sizes** of **Bacterial Cultures**

Some techniques are designed to determine the actual **population density**.
This involves working out the number or mass of **cells** in a particular volume of medium.

1) Direct **cell counts** are possible by taking a tiny sample of the medium and examining it under a **microscope**. You can then count the number of cells you see in a fixed volume of the medium using a **haemocytometer** (a grid of microscopic chambers, each with a known area and depth). It's best to do a number of counts and find the **average**, for it to be **accurate**. This number can then be **multiplied** to give the number in the sample.

0.2mm

how a haemocytometer would look through a light microscope

If this well has a depth of 0.1mm then the volume here is $0.2 \times 0.2 \times 0.1 = 0.004mm^3$ — this means there are about 5 cells per $0.004mm^3$, which is about 1250 cells per mm^3.

2) The **density** of cells can be measured with a **turbidometer** (colorimeter) — the more **turbid** (cloudy) the medium is, the less **light** is transmitted through the turbidometer. The turbidity is then **compared** to the turbidity of a **known** number of bacteria.
3) It's possible to measure **fungal** cultures by simply **weighing** them, but this isn't usually sensitive enough for bacteria.

The **Dilution Plating Method** tells us the Number of **Viable Cells**

All the methods above tell us the **total quantity** of cells in a sample of culture. The **dilution plating method**, however, can tell us the number of **viable** (living) **cells** in a medium — those that can divide. Here's how it's done:

1) A known volume of the culture is **diluted** by a known amount. It's then **further diluted**, again by a known amount each time.
2) A known volume of the final dilution is spread over an **agar plate** and **incubated**. Each single viable cell found in this final dilution will **reproduce** to produce a visible **colony**.
3) The number of colonies is **counted**. By knowing the **dilution factors** used each time, it is possible to work out the **total** number of cells in the first sample.

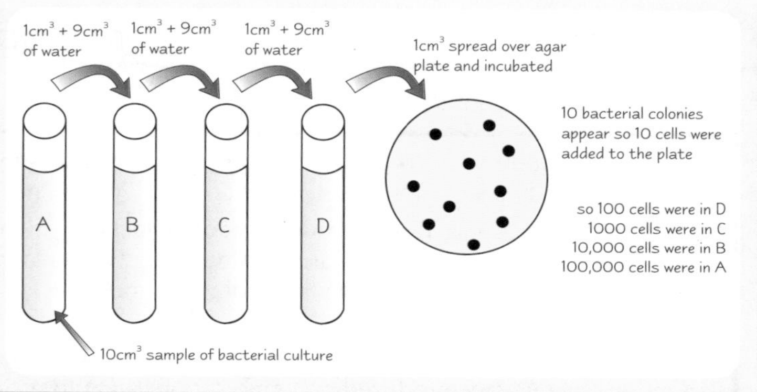

Many types of bacteria are potentially dangerous. This means that any labs growing bacteria must have built-in safety features that prevent contamination of workers and the environment. Regions of negative air pressure are created in microbiology cabinets, so that contaminated air from the lab flows into the cabinet through air flow hoods, instead of spreading around the environment.

Practice Questions

Q1 Which factors result in lag, exponential, stationary and decline phases in a bacterial population growth curve?

Q2 What is meant by the term 'secondary metabolite'? Give an example of a secondary metabolite.

Q3 Describe the basis for the turbidity technique in establishing the density of a bacterial culture.

Q4 Describe one safety feature of a microbiology lab.

Exam Questions

Q1 Outline how the dilution plating method could be used to compare the number of viable bacterial cells at two points in time. [8 marks]

Q2 a) Explain why a culture of bacteria in a nutrient broth showed increased population growth when oxygen was bubbled through the medium. [3 marks]

b) Suggest how the results would be different for a bacterial culture of an obligate anaerobe. [2 marks]

Graphs? OK, that's it — I want out...

Now, don't be scared off by those nasty-looking graphs. They're really not that bad — you've dealt with much worse in your GCSEs, so you can definitely handle them. And then the rest of the page is dead easy... You might even have played about with these ways of counting cells at school or college. And if not, then well, it's not that hard to learn anyway.

Industrial Growth of Microorganisms

The large-scale culturing of microorganisms means we can produce lots of important substances... like yoghurt. Great.

*Many **Industries** Grow Microorganisms on a **Large Scale***

All sorts of microorganisms can be grown to provide useful **products**. Many of these products are important in the **food industry**, such as alcohol in wine-making or lactic acid in yoghurt manufacture. Others are grown for **medical** applications, such as for antibiotic production.

These substances are produced in the metabolic reactions of the microorganisms.

Large cultures of microorganisms are grown in vessels called **industrial fermenters**. The culture methods used are designed to **increase** the growth rate of microorganisms through creating **optimum conditions** (see page 80). There are two main **culture methods** that are used:

Batch culture occurs in a **fixed volume** of medium. **Oxygen** is usually added during the growth of the microorganism and **waste gases** removed. Growth occurs up to the **stationary** phase until food becomes depleted and excretory products accumulate. Eventually **new cultures** must be established to start the process over again in a different batch. It is easier to **control** conditions in a batch, and **isn't costly** to start again if contamination occurs.

In **continuous culture**, fresh, sterile medium is added to the culture at a constant rate. Used-up medium and dead cells are removed at a constant rate. Cells are kept in the **exponential** growth phase and the culture lasts for far **longer** than with the batch method. This method is more **productive** than batch culture, and **smaller** vessels can be used. However, if **contamination** occurs, the whole lot will be lost, which is **very costly**.

*Conditions in a **Fermenter** are Controlled to **Optimise Microorganism Growth***

A typical **fermenter** consists of the following **components**:

Fermenters are also called bioreactors.

1) A motor drives the rotation of an **impeller**, which **stirs** the culture and continually brings fresh medium in contact with the microorganisms.
2) Cool water is circulated through a **cooling jacket**, which reduces the risk of **overheating**, especially in larger fermenters. Fermentation releases a great deal of heat, which is transferred to the cool water in the jacket by conduction and is carried away as the water flows out.
3) A **pH controller** contains a **pH probe** to detect pH changes (e.g. caused by acids produced by fermentation). It automatically delivers a **neutralising** quantity of **acid** or **base** when required. Constant pH is needed for optimum enzyme activity.
4) **Sterile air** is delivered through an **aerator** to provide **oxygen** to maximise aerobic respiration, and thus maximise growth.

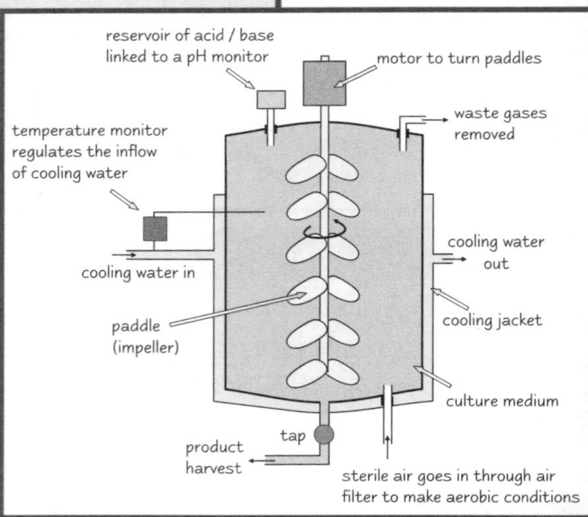

Large-scale fermenters are difficult to keep **sterile** and heating and cooling can be slower, making it more difficult to **control** conditions.

The whole fermenter is usually made of smooth stainless steel. This can be sterilised before use by pumping **steam** through the apparatus. Product harvest from a fermenter occurs through a **tap**.

After this, the product is **purified** and sometimes chemically **modified**. This series of treatments after harvesting is called **downstream processing**.

*Microorganisms are **Screened** for **Suitability***

Microorganisms are **screened** to maximise the production of certain substances. **Antibiotic production**, for example, can be detected in a Petri dish by placing samples of the antibiotic-producer with colonies of antibiotic-sensitive bacteria. Colonies disappear as they die in the vicinity of the antibiotic-producer.

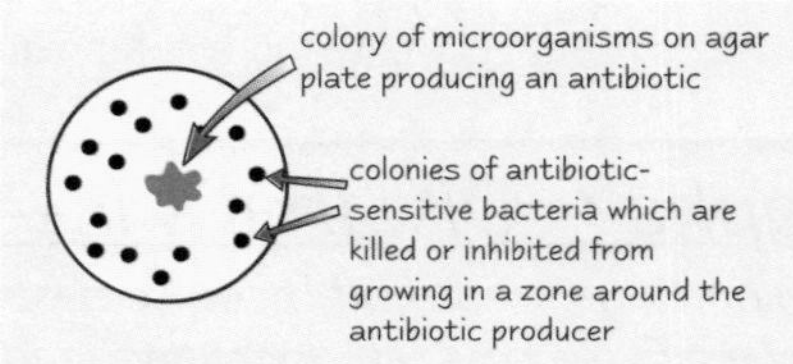

Microorganisms can also be screened for **enzyme production**. For example, **protease** production by bacteria can be detected by its ability to clear a cloudy suspension of the protein **casein**.

Industrial Growth of Microorganisms

Penicillin is Produced by a Specific Type of **Mould**

Penicillin is an **antibiotic** made by a mould called *Penicillium chrysogenum*. It's grown in **fermenters** by **batch culture** because it's a secondary metabolite and only starts to be produced in quantity once the exponential growth phase is over.

Here's how penicillin is made:

1) Organic **carbon** and **nitrogen** sources are added to the medium, as well as **oxygen** for aerobic respiration.
2) **Downstream processing** of the penicillin begins by **filtering** the mould mycelium from the medium.
3) The penicillin is **extracted** in **butylacetate** (an organic solvent).
4) It is **concentrated** and **crystallised** using **potassium salts**.
5) It may be chemically **modified** so that its effects can overcome penicillin-resistant strains of bacteria.

Enzymes and **Microorganisms** are used in **Industry**

Many industrial processes are **catalysed** by enzymes. Some of these processes use **isolated enzymes**, as opposed to using whole **microorganisms** for their enzymes. These are often **extracellular enzymes** — the ones that are naturally **secreted** from cells and so work outside of cells. As these enzymes are **naturally** secreted from cells, they don't need to be **extracted**, which is good as extracting enzymes can be complicated. Isolated enzymes are used in a wide variety of **industries**, including the production of food, textiles, leathers and medicines.

Some industrial processes (like **brewing**) use whole **microorganisms**. These microorganisms have **enzymes** that catalyse the reactions in the industrial processes. So it's really the enzymes, not the microorganisms themselves, that are useful.

Isolated enzymes are **easier** to use than microorganisms — it's easier to **isolate** and purify their products, and as only a single reaction is catalysed, it's easy to control **conditions** to optimise this.

Isolated Enzymes can be **Immobilised**

When enzymes are used in industrial processes, they end up **dissolved** in **solution** with their substrates and products. The **product** needs to be **separated** from this mixture, which can often be quite complicated.

Many industrial processes therefore use **immobilised enzymes**, which don't need to be separated out afterwards. These are enzymes attached to an **insoluble material**, which could be **fibres** (e.g. of collagen or cellulose) or **silica gel**, or they could be encapsulated in **alginate beads**. (Alginate is a jelly-like substance.) The substrate solution can be run through a **column** of **immobilised enzymes**. The active sites of the enzymes are still available to catalyse the conversion of substrate into product, but only **product**, not product and enzyme, emerges from the column.

This method has several **advantages**:

1) The insoluble material with attached enzymes can be washed and **re-used**. The enzyme molecules remain **active** and so don't need to be continually purified (which can itself be expensive).
2) The enzymes **don't contaminate** the product.
3) The immobilised enzymes are more **stable** and less likely to denature at high temperatures or extremes of pH.

Immobilised lactase is used industrially to break down the lactose in milk, so that dairy produce can be made safe for those who are lactose-intolerant.

Practice Questions

Q1 Describe how each of the following components of an industrial fermenter will optimise the growth rate of the cultured microorganism: water cooling jacket, pH controller, stirrer.

Q2 Explain what is meant by the immobilisation of enzymes.

Exam Question

Q1 a) Distinguish between the terms batch culture and continuous culture. [2 marks]

b) Give one advantage each for batch culture and continuous culture. [2 marks]

c) Explain why the antibiotic production of a *Penicillium* fermenter would decrease if the flow of sterile air into the reaction vessel were interrupted. [3 marks]

I'm well cultured, me...

You just wouldn't believe how important this stuff is. So many things we love and need depend on stuff that's grown in fermenters. Take Quorn for example. It's made from a lovely mould in continuous culture. So without fermenters, all you veggies wouldn't get to eat those lovely fake ham slices and fake meat pies, and fake chicken pieces. Mmmm... lucky you.

Bacterial Disease

Now you're going to learn all about nice bacterial diseases and how much they make you throw up, swell in funny places or, erm, excrete excessively. You're also going to learn how to avoid these lovely afflictions, should you so desire.

Some Types of Bacteria are Pathogenic

Some bacteria are **parasites** — they **invade** a living body and obtain **nourishment** from it. A **pathogen** is any microorganism that is a **parasite** of a plant or animal and causes **disease**. Whether bacteria are pathogenic depends largely on the structure of their **cell wall** and **capsule**. These have features that enable the bacteria to:

- invade the body of the **host**
- **attach** to the cells of the host
- some species can even invade the individual **cells** of the host

Bacteria that are better equipped to invade a host and cause a disease there are said to have high **pathogenicity**. The **number** of bacteria needed to cause a certain disease is an indication of the **infectivity** of the bacteria. The **invasiveness** is the ability of bacteria to **spread** within the host.

Here are examples of the **infectivity** of two types of bacteria that cause food poisoning:

- ***Salmonella typhi***, with **high infectivity** (around 0.1 to 1 million cells needed), causes **typhoid fever**.
- ***Salmonella enteritidis***, with **low infectivity** (around 1 to 10 million cells needed), causes **salmonellosis** (Salmonella food poisoning).

Many Pathogenic Bacteria Produce Harmful Toxins

Most kinds of pathogenic bacteria produce chemicals as **by-products** of their metabolism. Chemicals that can cause harm inside the body of the host are called **toxins**.

Exotoxins are soluble chemicals that are **released** from the cells. They are produced by bacteria such as ***Staphylococcus*** and ***Vibrio cholerae***, which causes cholera.

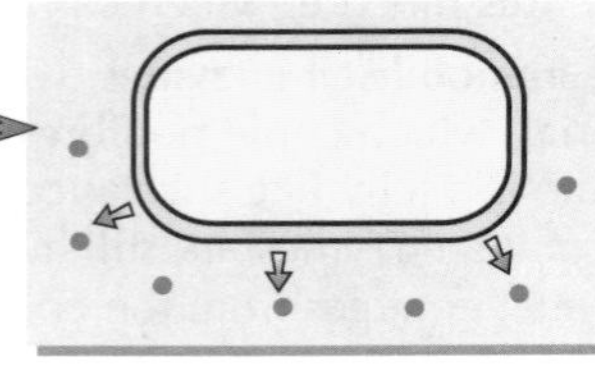

Exotoxins released from bacterium (e.g. Staphylococcus aureus) quickly cause food poisoning.

Endotoxins are **retained** inside the bacterial cells, but are released when the cell walls and membranes of these bacteria are damaged. ***Salmonella*** produces endotoxins.

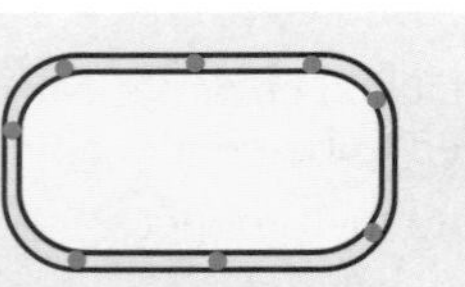

Endotoxins inside cell, released on death of bacterium (e.g. Salmonella enteritidis) cause food poisoning after a delay.

Usually **higher** concentrations of endotoxins than exotoxins are needed to produce the symptoms of disease.

Bacterial Toxins have Different Effects on the Body

The **effects** of bacterial toxins on the body range from vomiting to fever to swellings and other nasty stuff.

Food poisoning is caused by toxins from bacteria in food or water. The exotoxins of ***Staphylococcus***, for example, produce vomiting and diarrhoea within a few hours of eating contaminated food.

Salmonella enteritidis bacteria, on the other hand, occur in undercooked poultry. They stick to the gut epithelium where they are taken up by phagocytosis. They divide and produce endotoxins that cause inflammation and fever. Diarrhoea occurs a day or so after eating contaminated food.

Escherichia coli bacteria are found in food. Their exotoxins have a variety of effects on the body, from upset stomachs to urinary tract problems and septicaemia (blood poisoning).

Typhoid fever is also caused by a type of *Salmonella* bacterium.

Cholera is caused by *V. cholerae*, mainly found in water. The exotoxins cause diarrhoea.

Bacterial Disease

Try to **Avoid** Getting a **Bacterial Infection**

Here are a couple of tips on reducing your chance of getting a **bacterial disease**:

1) Food (especially poultry) must be thoroughly **cooked** to make it safe to eat. This **kills** bacteria, such as *Salmonella*. It's also important to make sure that frozen meat is properly **thawed** before cooking, or the centre might not reach the right **temperature** for the bacteria to be killed.
2) Many disease-causing bacteria, such as those causing typhoid, can live in water, and so can be transmitted by **drinking** from infected water supplies. Infection might happen if the water is **contaminated** by raw sewage, for example — something that often happens in less economically developed countries. It's therefore important to develop proper **sewage systems** that discharge downstream of drinking water sources. Sewage can also be **treated** to filter out and kill microorganisms.

Some individuals are **carriers** of disease-causing organisms — they **don't** show the **symptoms** of the disease. This could be because they're in the very **early** or **late** stages of the disease. Some individuals simply never develop the disease, but they could be **spreading** it without knowing.

Transient carriers only carry the pathogen for a brief time, but **chronic carriers** have it for months or years. Carriers of typhoid may continue to pose a threat in a population, even after all the sufferers have been cured.

Treating **Diarrhoea** Means Replenishing Lost **Body Fluids**

Diarrhoea is one of the most obvious symptoms of food poisoning. It quickly leads to **dehydration** and loss of **salts**. Treatment therefore consists of a form of **rehydration therapy** — consuming fluids that replace both lost water and lost salts.

Antibiotics rarely help in cases of food poisoning because the diarrhoea makes it difficult for the drug to be **absorbed** into the bloodstream from the gut cavity.

The diarrhoea sufferer's best friends.

Practice Questions

Q1 Distinguish between the terms pathogenicity, infectivity and invasiveness as they are applied to disease-causing bacteria.

Q2 Distinguish between exotoxins and endotoxins, naming examples of bacteria that produce each type.

Q3 Name two illnesses that *Escherichia coli* can cause.

Q4 Give two ways of preventing bacterial infection.

Exam Questions

Q1 Describe how the bacterium *Salmonella enteritidis* can bring about food poisoning. [4 marks]

Q2 Describe how *Staphylococcus* bacteria can bring about food poisoning. [3 marks]

Ooh, goody, a page on diarrhoea... Help me please...

*Not a happy couple of pages — nasty diseases, food poisoning, diarrhoea. Still, at least it's a bit more interesting than reading about the use of enzymes in industry (*yawn*). Here's a clever little tip for you — learn how to spell diarrhoea before your exam. You'll only waste time worrying and crossing out different spellings otherwise. So — D-I-A-R-R-H-O-E-A.*

Viral Disease

Viruses are acellular microorganisms, so they're not cells like bacteria. They're not even living things. So they don't work like bacteria. So don't ever, ever, say they do. Or I'll come round your house with a big stick and a vivid imagination.

All Viruses **Invade** Living Organisms

Here's a bit about the basic **structure** and **workings** of **viruses**.

1) Viruses only become **metabolically active** when they've **invaded** a living cell.
2) Because they invade cells, viruses are **pathogenic** — they can cause **disease**.
3) Viral particles are called **virions**. Most are less than 0.2 μm in diameter, so they're only visible through an **electron microscope**.
4) All viruses contain **nucleic acid** (either DNA or RNA). These are the molecules that replicate in order for the viruses to **reproduce**.
5) Each particle is surrounded by a protein coat called a **capsid**, composed of protein units called **capsomeres**.
6) Some viruses, in addition, are surrounded by another layer of **lipid** called the **envelope**. This is made from the **cell membrane** of the host cell that the virus invades.

Viruses **Replicate** Inside their **Host Cells**

Viruses can only **reproduce** and make **protein** inside a **host** cell. They don't have the equipment, such as **enzymes** and **ribosomes**, for doing this on their own, so they use those of the **host**. The **life cycle** of a virus is an '**infection cycle**':

1) The capsid or envelope of the virus particle **attaches** to a **receptor** molecule on the host cell membrane.
2) The virus **penetrates** the cell and the capsid **uncoats** to release the **nucleic acid** into the host cell's cytoplasm.
3) The nucleic acid **replicates** and is **transcribed** to make **mRNA**. **Translation** occurs on the host's **ribosomes** to make more capsid **proteins**.
4) New viruses are **assembled**. Then they are **released** from the cell by **lysis** (splitting) of the cell or by **budding** from the cell.

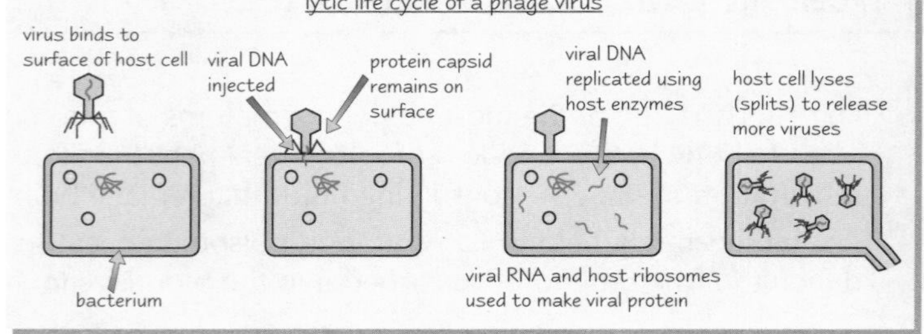

HIV is the Virus which leads to **AIDS**

Human immunodeficiency virus (HIV) affects the white blood cells known as **helper T-cells** in the immune system. It eventually leads to **acquired immune deficiency syndrome (AIDS)**. AIDS is a condition where the immune response worsens and eventually fails. This makes the sufferer more **vulnerable** to other infections, such as pneumonia.

HIV is a **retrovirus**. This means it carries an enzyme called **reverse transcriptase**, which does this:

1) Inside the white blood cell, reverse transcriptase is used to make a **complementary strand** of DNA from the **viral RNA template**.
2) From this, **double-stranded DNA** is made, which is **inserted** into the human DNA.
3) The DNA stays in the white blood cell for a long while (the **latency period**).
4) Eventually, after a long period of time, the virus particles **multiply**. The DNA **codes** for the synthesis of **viral proteins** to make new viruses that burst out, killing the cell, and then **infect** and **destroy** other white blood cells.

protein
reverse transcriptase
lipid envelope
RNA
Human Immunodeficiency Virus

The **Flu** Virus also Contains **RNA**

The **influenza virus** infects **epithelial cells** of the nose, mouth, throat, trachea and bronchi. The **lipid envelope** of the virus fuses with the **cell membrane** of the host, so that it can penetrate the **cytoplasm**, and then the viral RNA enters the **nucleus**. Here it uses its own **viral RNA polymerase** to replicate the **viral RNA**. Then influenza proteins are made on host **ribosomes**. The assembled viruses are released from the dying cell by **exocytosis**. The accumulation of dead cells makes **mucus** build up — the most obvious symptom of flu. (Remember that, unlike HIV, the flu virus **doesn't** have reverse transcriptase.)

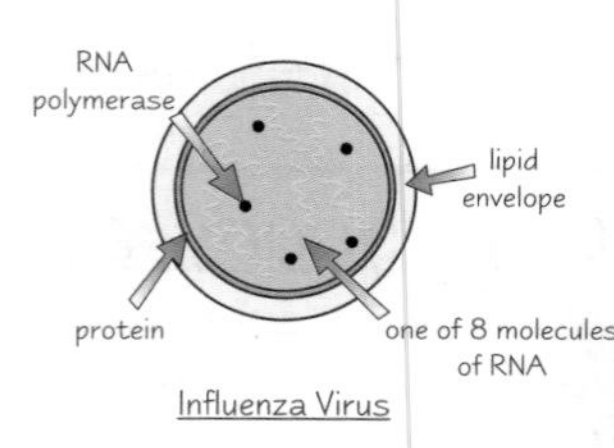

Viral Disease

It's **Difficult** to **Treat** Viral Diseases

Diseases caused by viruses are **difficult** to **treat** because the viral pathogens are actually **inside** the host cells. Some actually **incorporate** their nucleic acid into the host DNA too, which means that the host cell is able to **produce** even more viruses. Here are a few ways of dealing with viral diseases like flu:

1) **Drugs** used to treat viral diseases are designed to target infected cells in a similar way to special white blood cells called **killer cells** (a type of T-lymphocyte). These actively **seek out** and **destroy** infected host cells by **lysis**. They recognise the **protein coat molecules** of the virus that were left behind when the virus invaded the cell. Other types of drugs act as **inhibitors** of specific viral **enzymes**, such as reverse transcriptase.
2) Treatment of viral diseases is more often about **relieving** the **symptoms** than destroying the virus. Apart from mucus production, the symptoms of **flu** are fever, headache and muscular pain, which can be treated by **painkillers** (e.g. aspirin or paracetamol). It's important to drink a lot of **fluids** too to replace those that are lost.
3) Flu is **infectious** during the time you're showing symptoms, and for about a week afterwards. This means that its **spread** can be controlled by **isolation** of sufferers, although that's not much fun.
4) **Vaccination** against flu is possible, but the vaccine must be based on many **antigens** as the virus **mutates** a lot.

Viruses are **Transmitted** in Different Ways

Given the **difficulty** of treating viral diseases with **drugs**, the best way to control them is by reducing their **spread**. Viral particles can be **transmitted** in a variety of ways, depending upon the type of virus:

1) Some viruses, such as **flu**, are transmitted through **air** and so are contracted via the **respiratory system**. This is called **droplet infection**.
2) Some, such as **HIV** or **herpes**, are transmitted by **sexual intercourse** or **blood transfusions**.
3) Others, such as **yellow fever**, are transmitted by **biting insects**.
4) A few, such as **hepatitis A**, are transmitted via contaminated **food** or **water**.

This means that different **control strategies** are needed depending on how the virus is **transmitted**.

Transmission of **HIV**, for example, can be reduced by using **barrier contraceptives** like condoms, heat-treating blood products and not sharing hypodermic needles. Transmission of viruses spread by **biting insects** might be reduced with **insecticide**.

The main way of reducing the spread of viruses, though, is simply to **isolate** the sufferer in **quarantine**, so they can't pass it on.

Practice Questions

Q1 Describe the lytic life cycle of a virus.

Q2 HIV is a retrovirus. What does this mean?

Q3 Explain how infection with HIV can lead to the development of AIDS.

Q4 Explain why most viral infections are more difficult to treat than bacterial infections.

Exam Questions

Q1 Explain what is meant by the term 'latency period' with reference to HIV infection. [2 marks]

Q2 AZT (azidothymidine) is an example of a drug that has been shown to prolong survival of AIDS sufferers. It is thought to be an inhibitor of reverse transcriptase. Explain how this effect reduces the number of viral particles that can be released from infected cells. [3 marks]

You can't avoid all viruses, so just learn to love 'em. Little darlings.

You see that question on AIDS drugs? Well, don't take that as evidence that there's a cure for AIDS, cos there's not. All the AIDS drugs do is prolong the life of sufferers. Although, obviously, that's a massive step in the race to find a cure, so it's not to be scoffed at. I'm just reminding you not to write that there's a cure for AIDS, cos, I've said it before — there's not. Yet.

Protection Against Disease

You'll know a lot about white blood cells and their disease-fighting friends from your earlier studies. This stuff's just the same really, but with more long words to learn. Lucky you.

The First Line of Defence Stops Pathogens from Infecting the Body

1) The **skin** acts as a barrier, preventing the **entry** of microorganisms. The outermost layers make up the **epidermis**, where the cells closer to the surface are strengthened by the protein **keratin**. **Sebaceous glands** of the lower **dermis** layers produce water-proofing oils that also keep hair follicles relatively free of **bacteria**.
2) **Sweat**, **tears** and **saliva** contain an enzyme called **lysozyme**, which causes the **lysis** (splitting) of bacteria, and prevents them from infecting the living cells of the skin, respiratory and digestive tracts.
3) **Mucus** produced in the **respiratory tract** can trap bacteria and other microorganisms, preventing further invasion of the body. Mucus also contains **lysozyme**.
4) **Cilia** lining the trachea beat steadily, causing a net flow of **mucus** (with its entrapped microorganisms) towards the back of the throat, where they are **swallowed**. This prevents microorganisms from reaching the **alveoli**.
5) The concentrated **hydrochloric acid** of the **stomach** denatures the enzymes of microorganisms. **Pepsin** in the stomach also **hydrolyses** them (breaks them down by reacting them with water).

White Blood Cells of the Immune System Destroy Pathogens

There are loads of types of **leucocytes (white blood cells)** that defend the body against the microorganisms which manage to get through the first line of defence. These leucocytes are all produced in the **bone marrow**. Some kinds, though, only become functional once they've passed through special **lymphoid tissue**, such as the **thymus gland**.

Most leucocytes are **granulocytes**, so called because they have a granular cytoplasm due to the presence of **vesicles**, such as lysosomes. Here are three types of **granulocytes**:

- **Neutrophils** (about 70% of leucocytes) are **phagocytic**.
- **Eosinophils** (1.5%) have **anti-histamine** effects.
- **Basophils** (0.5%) produce **histamine**. (**Histamine** is the substance that makes blood capillaries leaky during **inflammation**.)

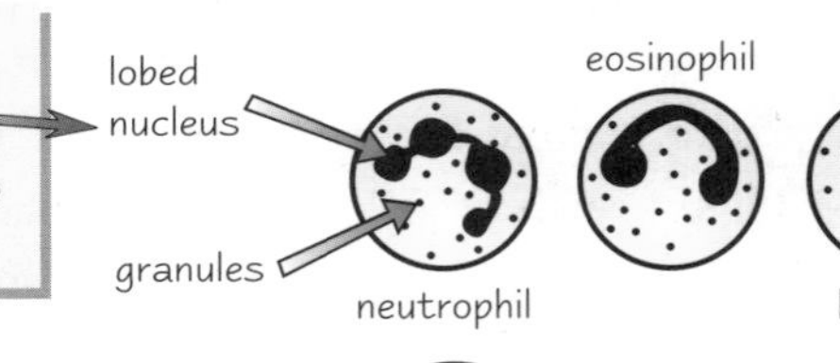

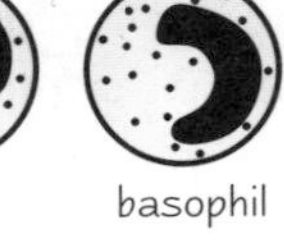

Other leucocytes are **agranulocytes**:

- **Monocytes** (4%) are **phagocytic**.
- **Lymphocytes** (24%) are involved in antibody production (**B-lymphocytes**) and cell-mediated immune responses (**T-lymphocytes**).

Phagocytosis is where Phagocytes Destroy Invading Pathogens

Some white blood cells, **neutrophils** and **macrophages** (monocytes that are found in tissue fluid), are **phagocytic**. Here's what they do:

1) **Local** infections of bacteria cause **basophils** to produce **histamine**, which brings about **inflammation**.
2) Inflammation means that blood capillaries become dilated and **leaky**, allowing **phagocytes** to leave the blood and reach the **site** of infection. (Phagocytes squeezing through capillary wall pores is known as **diapedesis**.)
3) Phagocytes **engulf** microorganisms like bacteria, enclosing them within **food vacuoles**. **Lysosomes** inside the phagocytes **fuse** with the food vacuole. These contain **hydrolytic enzymes** to break down the microorganism.

Lymphocytes have Various Immune Responses

Each type of lymphocyte attacks a **specific** type of foreign particle (**antigen**). Lymphocytes belong to several classes:

1) **Killer T-lymphocytes (cytotoxic T-lymphocytes)** bind to cells (e.g. those infected by viruses) and kill them by **lysis**.
2) **Helper T-lymphocytes** stimulate **B-lymphocytes** to **divide** into antibody-producing plasma cells. B-lymphocytes can't work without them.
3) **Plasma B-lymphocytes** secrete **antibodies** into blood plasma.
4) **Memory B and T-lymphocytes** remain in the blood for many years and produce a more effective **secondary immune response** to an infection, which is the basis for **long-term immunity**. See pages 90-91 for more details.

Protection Against Disease

Invading Foreign Particles Stimulate the Body's Defence Response

If the body is invaded by foreign particles, the body's **defence system** is activated. **Phagocytes** are the first line of defence, but a more **long-term** defence system is the production of **antibodies** — which leads to **immunity** to certain diseases.

1) The body defends itself when invaded by pathogens by setting up an **immune response**.
2) An **immune response** is a cellular response to the presence of an **antigen** (see below). Its purpose is to remove the antigens or make them **harmless**.
3) An **antigen** is a particle that is detected by the immune system as being **foreign**, or '**non-self**'.
4) Cells and molecules that are part of the body are described as **self**, since the immune system has become desensitised to them. Antigens are therefore described as **non-self** particles.
5) If non-self antigens are detected, the **helper T-cells** are activated and secrete **cytokines** to stimulate the appropriate **B-cell** (one that produces antibodies with a shape complementary to that of the antigen) to divide by mitosis and form plasma and memory cells. The **antibodies** then seek out and **destroy** the foreign cells, or coat them and make it easier for phagocytes to ingest them.

Antigens Trigger Immune Responses

1) Don't get confused about what antigens are — whole microorganisms **aren't** antigens. Antigens are specific **molecules** — normally **proteins**, **polysaccharides** or **glycoproteins**. They are found on the **cell walls** of bacteria or the **capsids** of viruses, for example.
2) However, some **inorganic** substances, such as nickel, can act as antigens. This is why some people develop **allergies** to metal earrings.
3) Foreign invaders are normally **microorganisms**, so it's definitely a **good** thing that the immune system destroys them — it means you can fight off the diseases they cause.
4) Unfortunately, the immune system responds in exactly the same way when the body receives a **transplanted organ**. **Rejection** of the organ is simply a natural immune response to the **foreign antigens** on the cell surface membranes of the organ.

The type of immunity described above is known as **active immunity**. The other main type of immunity is **passive immunity**, which is **temporary**. Babies are **naturally passively immune** to a number of diseases when they're born, because some of their **mother's** antibodies have come across the placenta or in her breast milk. They have no '**memory cells**' though, so after a while the antibodies **break down** and the immunity goes. The passive immunity protects the baby in its first few months, and gives it a chance to build up its own **active immunity** to common pathogens.
(See the next couple of pages for a bit more on how an immune response actually happens.)

Practice Questions

Q1 Describe three ways in which the body prevents entry of microorganisms.

Q2 Describe the process of phagocytosis.

Q3 Where are antigens usually found?

Q4 Name two examples of an antigen.

Exam Questions

Q1 Explain the role of lysosomes in phagocytosis. [3 marks]

Q2 Describe the process of diapedesis. [2 marks]

Q3 Define the term 'antigen'. [3 marks]

My Auntie Jen doesn't get on with my Auntie Biddy...

Phew — so many big words on this page. Never mind, it shouldn't be too hard to fit them all in your head. You know a lot about white blood cells and immunity already from AS level and even your GCSEs... so it's just a case of adding a few more words into that big bank of knowledge in your head. I'd learn how to spell those long words too, if I were you.

Cell and Antibody Mediated Immunity

This bit's all about the two types of immune response — the one with B-lymphocytes, and the one with T-lymphocytes.

There are Two Types of Immune Response

Remember that B-lymphocytes are to do with blood, and T-lymphocytes are to do with tissues. Easy eh?

The two types of immune response are the **humoral** and **cell-mediated** responses. **Humoral** responses involve **B-lymphocytes**, and **cell-mediated** responses involve **T-lymphocytes**. Here's how these different types of lymphocytes are **made** and start to **work**:

1) Both types of lymphocyte involved are made in the **bone marrow**.
2) However, **T-lymphocytes** need to pass to the **thymus gland** (in the chest cavity) during infancy to start working.
3) **B-lymphocytes** mature in the **bone marrow**, **lymph nodes**, and **foetal** (but not adult) **liver**.
4) When lymphocytes have become functional, they are said to be '**competent**'.

B-Lymphocytes Release Antibodies into Blood Plasma

1) The **humoral response** (also called the **antibody-mediated** response) occurs when the presence of an antigen stimulates **B-lymphocytes** to produce and secrete **antibodies** into the blood **plasma**.
2) The antibodies circulate in the bloodstream and **bind** to specific **antigens**.
3) When this happens, the antigen-carrying particle is made **harmless** in some way (see below).

Antibodies are also called immunoglobulins.

Antibodies are **Y-shaped proteins** with a **quaternary structure**. This means that each molecule consists of more than one **polypeptide chain**. In fact, each has **four** chains joined by **disulphide bridges**.

Each type of **antibody** has specific **antigen-binding sites** with a specific **amino acid sequence**. This part of the antibody **binds** to the antigen, in the same way that the active site of an enzyme binds to a substrate.

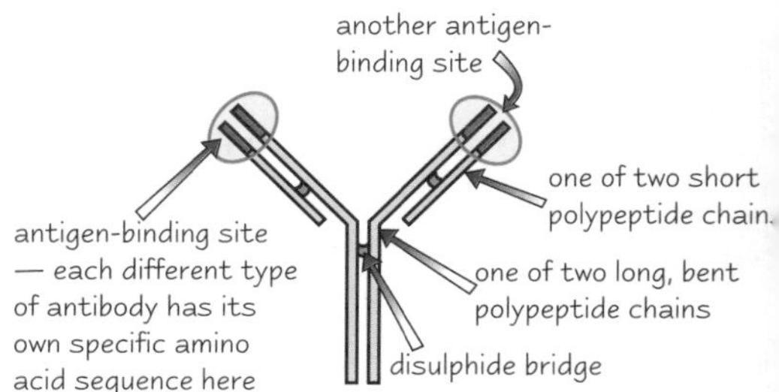

Antibodies make Antigens Harmless

When an antibody **binds** to an antigen, forming an **antibody-antigen complex**, it brings about a change that makes the antigen **harmless**. This works in one of several ways:

1) **Agglutination** happens — the antibodies stick to the surface of the microorganisms, making them **clump** together and rendering them harmless. (This also happens in incompatible **blood transfusions** — antibodies in the recipient's blood respond to antigens in the donated blood, making donated blood agglutinate.)
2) The antibodies sticking to the surface of the microorganism cells can also sometimes bring about **lysis** (breaking up) of the cells.
3) **Phagocytosis** by phagocytic white blood cells (e.g. neutrophils) is stimulated or speeded up.
4) **Precipitation** happens to soluble toxins — the toxins, which were originally **dissolved** in solution, become **solid** and therefore can't do any harm.
5) **Bacteria** are **prevented** from attaching to host cell membranes.

T-Lymphocytes have Receptors that Bind to Antigens

The **cell-mediated response** involves **killer T-lymphocytes**. T-lymphocytes have **receptors** on their surface which recognise and bind with particular **antigens** on the surface of other cells (in a similar way to how **antibodies** work). So instead of producing antibodies to attack pathogens, killer T-lymphocytes bind to them **directly**. Killer T-lymphocytes can then **destroy** the foreign cell. Some T-lymphocytes have receptors that bind to cells infected by **viruses**, others destroy **cancer cells**. They bind specifically to the viral molecules placed on the cell surface membrane of the cell. The **virus-infected cells** are then destroyed. T-lymphocytes are associated with immune responses in **tissues** rather than in blood.

Cell and Antibody Mediated Immunity

New Antigens Trigger the Primary Immune Response

Lymphocytes are **activated** by the presence of **antigens**. This is known as '**clonal selection**' because the antigen causes the **selection** of an appropriate type of lymphocyte for **cloning**. If the body **hasn't** been exposed to a particular kind of antigen before, then a **primary immune response** is set up by the lymphocytes:

1) **Exposure** to a specific antigen causes an antibody or a receptor on a T-cell to **bind** with it. At this early stage, both the antibodies and receptors are **attached** to lymphocyte cell membranes.
2) This binding stimulates the lymphocytes to undergo **mitosis**, producing **clones** of themselves.
3) **B-cell** clones produce the **specific** type of **antibody** to bind to the **specific** antigen present. **T-cell** clones have the **specific receptors** on their surface.
4) Many of the **B-lymphocytes** then become **plasma cells**. This means that they **secrete** their antibodies into the blood plasma. They only last for a few days. The **T-lymphocyte** clones just retain their **receptors**.
5) The antibodies bind to the rest of the antigens away from the cell, producing the **primary immune response**.

A Second Infection Produces a Secondary Immune Response

Not all the lymphocytes stimulated during the primary immune response become plasma cells. Some become **memory cells** instead. These **don't** secrete their antibodies during the first infection, but instead stay **dormant** in the blood. The plasma cells **die** once the infection is over, but the memory cells don't. If there is ever **another infection** by microorganisms with the same antigens, the memory cells can respond **immediately** by producing **antibodies** and **dividing** to create more plasma cells.

This is known as a **secondary immune response**. This time they secrete **more** antibodies, and do it more **quickly**. This secondary immune response is therefore **greater** and **faster**.

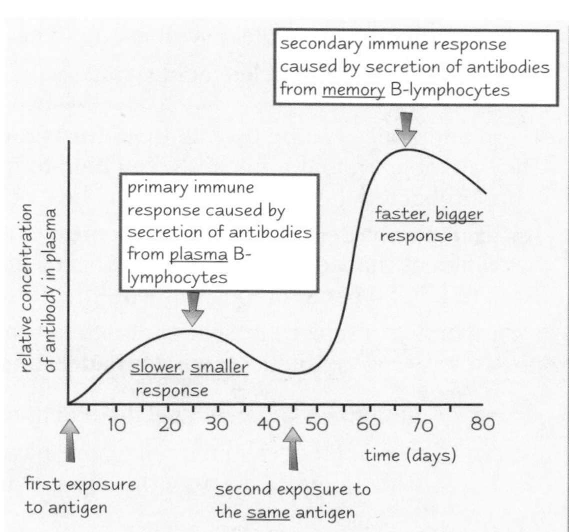

Genes occasionally **mutate** and, when they do, this could affect the **shape** of the antigens that are produced. This is known as **antigenic variability**. It means that the **antibodies** produced by a host body might not **bind** to the antigens anymore, and so an immune response might not happen.

Pathogens that mutate a lot can therefore do a lot of **damage** to their host. There are, for example, many mutated forms of the **flu virus**, which makes it really hard for the body to develop **immunity** to flu.

Practice Questions

Q1 Describe the structure of an antibody molecule.

Q2 Describe the difference in function between B-lymphocytes and T-lymphocytes.

Q3 Which is faster, the primary or secondary immune response?

Exam Question

Q1 a) With reference to the clonal selection theory, explain how an antibody-mediated immune response is set up when a body is invaded by a specific type of antigen. [5 marks]

b) Explain why antibodies don't bind to other molecules that are naturally found in the body. [1 mark]

Wish you could develop immunity to revision...?

Urgh, I bet you're ready for a nap now, aren't you? Lymphocyte this and lymphocyte that. Antibody this, antigen that. Zzzzzz. Well, I'm afraid there's more of it to come. Still two more pages on this topic before you get a new one. Sorry. But, anyway, here's an interesting fact — oh, erm, actually I can't think of one. Sorry again. Oops, I'm sorrying. Sorry.

Antibiotics and Vaccines

These pages are all about antibiotics and vaccines — what they do and where they come from. There's also a bit about antibiotic resistance, which is quite scary — the evolution of super-bacteria which aren't killed off by antibiotics. Eeek.

Antibiotics Kill or Prevent the Growth of Bacteria and Fungi

Antibiotics are **chemicals** produced naturally by a range of different microorganisms (especially fungi and bacteria). They **kill** or inhibit the **growth** of other types of microorganisms.

- **Microbicidal** antibiotics **kill** microorganisms.
- **Microbistatic** antibiotics inhibit **growth**.

Antibiotics are **secondary products** of the **metabolic reactions** of a microorganism, and are released from their cells. Although the microorganism uses up **energy** producing the antibiotic, it is an **advantage** to do so because antibiotics prevent the growth of other kinds of microorganisms. This reduces the numbers of **competitors** for food.

Antibiotics are often very **specific** in their mode of action. They usually work by acting as **inhibitors** of enzymes and other types of proteins. Some antibiotics are so **damaging** to the affected cells that the cells are **killed**.

Antibiotics Inhibit Enzymes Involved in Crucial Metabolic Reactions

The reason why many antibiotics are so **effective** at reducing growth or killing microorganisms is because they **inhibit** the metabolic reactions that are **crucial** to the growth and life of the cell.

1) Some inhibit enzymes that are needed to make the chemical **bonds** in bacterial **cell walls** and **cell membranes**. This stops the cells from **growing** properly. An example is **penicillin**.
2) Some inhibit **protein synthesis** by binding to bacterial **ribosomes**. Examples are **tetracycline** and **streptomycin**.
3) Some inhibit **nucleic acid synthesis** by binding to bacterial **DNA** or **RNA polymerase**. An example is **rifampicin**.

These antibiotics can be used to treat **diseases** caused by bacteria. They are able to do this because they bind to molecules that are **only** found in **bacterial** cells, rather than the host cells.

For example, bacteria have **70S ribosomes**, whereas the eukaryotic host has different shaped **80S ribosomes**. This means that the antibiotic will work to **kill** the **bacteria**, but will leave the host cells alone.

It's important to **select** the right antibiotic for the treatment of **specific** diseases. Some antibiotics have a **broader range** of effects than others:

antibiotic action targets on a bacterial cell

antibiotic could bind t enzyme needed to buil cell wall — cell would stop growing properly and could lyse (split)

antibiotic could bind to DNA or RNA polymerase, stopping DNA replication or transcription

antibiotic could bind to 70S ribosome, inhibiting protein synthesis

- Antibiotics that target **specific** reactions in **specific** microorganisms are called **narrow spectrum antibiotics**.
- Other antibiotics inhibit enzymes that occur in a **wide range** of microorganisms, and so can be used to treat a **wide range** of different disease-causing pathogens. These are called **broad spectrum antibiotics**.

Mutated Genes Cause Antibiotic Resistance

Microorganisms that are resistant to antibiotics have **mutated genes** causing a change in their chemical makeup. The antibiotic is prevented from **binding** to the enzyme, but the reason why depends on the type of **mutation**.

For example, bacteria that are resistant to **penicillin** have a gene causing them to produce an enzyme called **penicillinase** This **breaks down** the penicillin molecules before they can inhibit the enzyme that builds the bacterial cell wall.

Bacteria reproduce very quickly, and every time the cells divide, the DNA replicates. **Natural selection** causes the incidence of antibiotic resistance to rise quickly. This is because exposure of a population to an antibiotic **kills** those bacteria **without** the antibiotic resistance gene. Then the **resistant** ones, which obviously survive, can reproduce without **competition**.

Genes for resistance can also spread because of the exchange of **plasmids**. Plasmids are small **rings of DNA** in bacterial cells, and many contain antibiotic-resistance genes. Plasmids are exchanged when two bacteria join together in a process called **conjugation**.

Overuse of antibiotics has resulted in **antibiotic resistance** being a big problem in medicine. There are currently certain strains of disease-causing bacteria (e.g. the *Mycobacterium* that causes tuberculosis) that have **evolved** resistance to most common antibiotics. This makes the treatment of these diseases increasingly difficult.

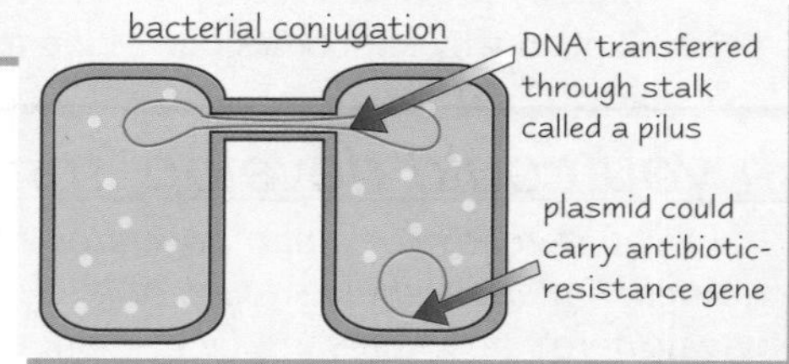

Antibiotics and Vaccines

Vaccines help out our *Immune Systems*

While your B-lymphocytes are busy dividing to build up their numbers to deal with a pathogen, you suffer from the disease. **Vaccination** can help avoid this. It lets your body increase its level of the B-lymphocyte and antibody needed to fight a particular disease, **without** the microbe being present and increasing in numbers at the same time. This means you get the **immunity** without getting any **symptoms**. Vaccines come in various types:

1) **Dead virulent pathogens** (e.g. **whooping cough**). The pathogen is **killed** but the antigens remain on its surface, so the body undergoes an immune response without the microorganism causing any damage.
2) **Live non-virulent strains** of a pathogen (e.g. **rubella**). Some strains of a pathogen stimulate **antibody production** but don't actually cause the disease.
3) **Modified toxins** (e.g. **diptheria**). Toxins (poisons) produced by pathogens can act as **antigens**. The toxin can be treated with **heat** or **chemicals** so that it doesn't produce any symptoms, but still triggers an immune response.
4) **Isolated antigens** (e.g. **influenza**). Sometimes, the antigen can be **separated** from the pathogen and injected to trigger an immune response.
5) **Genetically engineered antigens** (e.g. **hepatitis B**). The hepatitis B antigens have been **isolated** and can be made by **genetic engineering**. They're injected without the virus, so the disease can't develop.

Vaccines give you **active artificial immunity**. It's **active** immunity because your body makes its own antibodies, and will produce memory cells to help prevent the disease in the future. And it's called **artificial** immunity because you haven't actually **caught** the disease in the natural way — your body has been 'tricked' into reacting against a harmless pathogen.

Sometimes people need to be given '**instant immunity**' (vaccination takes a while to be effective). For example, if you get a bad cut you're often immunised against **tetanus**. This involves injecting the patient with the **tetanus antibodies** themselves, rather than with a harmless pathogen. It gives you **artificial** immunity **immediately**, but it's also **passive** immunity because your B-lymphocytes haven't been involved in producing the antibodies and there will be no memory cells. The immunity will wear off when the antibodies break down.

Vaccines not only protect the people vaccinated, but, because they reduce the **occurrence** of the disease, those **not** vaccinated are also less likely to catch it. This is called the **herd immunity effect**.

Practice Questions

Q1 Distinguish between the effects of the antibiotics penicillin, tetracycline and rifampicin.

Q2 What type of blood cell produces antibodies?

Q3 What is meant by the 'herd immunity effect'?

Exam Question

Q1 The graph on the right shows the number of cases of polio reported each year since 1991 in the Indian province of Tamil Nadu. A vaccination programme was in place throughout this period.

a) The same vaccine programme was in place over the years 1991-1993, yet the cases of polio dropped dramatically. Suggest a reason for this. [1 mark]

b) There are now very few cases of polio in Tamil Nadu. Explain why the vaccination programme should **not** be discontinued. [1 mark]

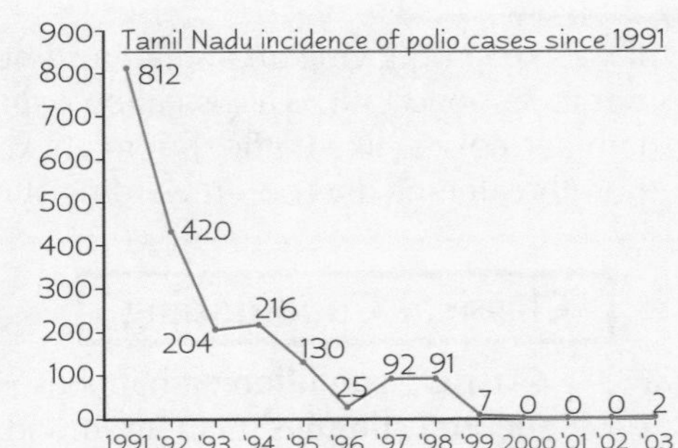

Q2 Explain how resistance to penicillin can spread through a population of bacteria. [4 marks]

Immunity — another good idea from God / random mutations / aliens...

Delete as appropriate, according to your own belief system. I'm a big believer in political correctness, me. Personally I think it was the aliens, but I'm not saying that's right, it's just my opinion. It's also my opinion that you need to get this page learnt, as thoroughly and as quickly as you can. See, some of my opinions are pretty sensible. Others — not so much.

Simple Behaviour Patterns

This section is for Option Module 8 — Behaviour and Populations.
This stuff is really interesting. It's the sort of thing you never knew you wanted to know about. You'll be inspired into watching documentaries on pigs learning to play computer games, and then end up doing a Psychology degree. Oops.

Behaviour is an Organism's Response to Changes in its Environment

Behaviour lets an organism **respond** and **survive**. Behaviour can either be **innate** or **learned**, but most behaviour relies on a blend of the two. Both your **genes** and your **environment** play a part in influencing your behaviour and it's sometimes hard to decide what is innate and what is learned. One example is **human speech** — almost all humans are born with the **innate ability** to speak, but if a child is brought up in **isolation** and doesn't hear people **speaking** for the whole of its childhood, then it won't ever learn to speak. Human babies have to **learn** language.

Innate Behaviour is Inherited and Instinctive

Innate behaviour is **instinctive**. Animals can respond in the **right way** to the stimulus **straight away**, even though they've never done it before, e.g. newborn mammals have an instinct to suckle from their mothers. Instinctive behaviour can be a fairly simple **reflex**, or a set of **complicated** behaviour, like a courtship ritual.

Bradley thought that the piglets' suckling instinct was getting rather out of control

Reflex actions are simple innate behaviours, where a stimulus produces a fairly simple response, like sneezing, **salivation**, coughing, and blinking. They often protect us from **dangerous stimuli**. Although reflexes are **involuntary** (automatic) actions, they can sometimes be **modified**, e.g. humans usually learn to **control** their bladder **sphincter** when they're quite young (see page 31).

Taxes and **kineses** are **reflexes** which allow animals to move away from **unpleasant stimuli**.

1) A **taxis** is a **directional movement** in response to a **stimulus**, e.g. earthworms show **negative phototaxis** — they move away from light.
2) A **kinesis** is a change in the rate of movement, depending on the intensity of a stimulus, e.g. **sea anemones** wave their tentacles more when stimulated by **chemicals** emitted by their prey.

Innate responses are often stereotyped. **Stereotyped responses** are controlled by **sequences** of behaviour called **Fixed Action Patterns** (**FAP**), e.g. nest building, courtship. These are **automatic** and the **same type** of **response** is always given to a stimulus, with **fixed patterns** of **coordinated** movements.

Learned Behaviour is Modified in response to Experience

Learned behaviour isn't **innate** — you have to learn it, obviously. It lets animals **respond** to **changing conditions**. Animals can learn to avoid predators and harmful food, and to find food or a suitable mate. Some important examples of learned behaviour are **habituation**, **classical conditioning**, **operant conditioning** and **insight behaviour**:

Habituation

If you keep on giving an animal a stimulus that isn't **beneficial** or **harmful** to it, it quickly learns **not** to respond to it. This is called **habituation**. That's why you can often **sleep through** loud and familiar noises, like traffic, but might wake up instantly at a quiet but **unfamiliar** noise. By **ignoring** non-threatening and non-rewarding stimuli, animals can spend their time and energy more **efficiently**.

Classical Conditioning

Classical conditioning happens when an animal **learns** (passively) to associate a '**neutral stimulus**' with an important one, e.g. a dog associates a bell ringing with the arrival of food. The response is **involuntary**, **temporary** and **reinforced** by **repetition**.

Ivan Pavlov — Classical Conditioning in Dogs

Pavlov studied the behaviour of dogs and noticed that they would **salivate** (drool) every time they saw or smelt food. He began to ring a **bell** just before each time the dogs were given their food. After a while he found that the dogs salivated when the bell was rung even if he didn't give them their food.

Simple Behaviour Patterns

Operant Conditioning

Operant conditioning or **'trial and error learning'** is where an animal **learns actively** to **associate** an **action** with a **reward** or a **punishment**. This happens in **humans** when children are rewarded or punished for **specific behaviour**.

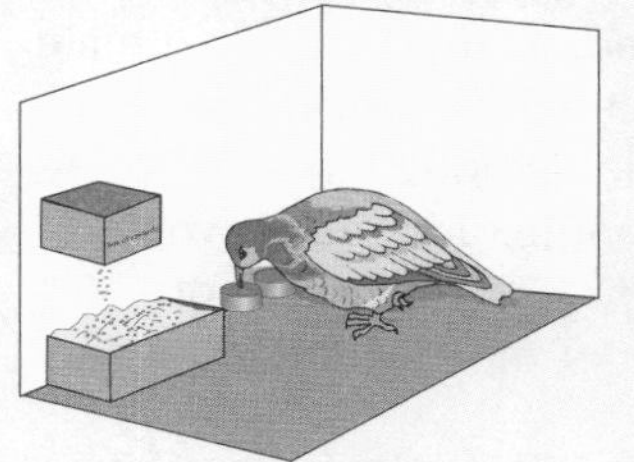

Burrhus Skinner — Operant Behaviour in Pigeons and Rats

Skinner trained **rats** and **pigeons** to obtain a **food reward** with a **'Skinner box'** that he invented. The animal had a choice of buttons to press. When the animal pressed a particular lever or button, it was **rewarded** with food. He found that pigeons and rats used a system of **trial and error** to learn which button to press to get the reward.

Insight Behaviour is the most Complex form of Learned behaviour

This type of behaviour involves **solving a problem** by looking at it, thinking about it, and using **previous experiences** to help solve it:

Wolfgang Kohler — Insight Behaviour in Chimpanzees

People used to think that only humans made and used tools. **Kohler** worked with **chimpanzees**, putting them in a play area with, for example, a bunch of **bananas out of reach**. To get the bananas, the chimp would have to **use** an **object** as a **tool**. The chimp had different length sticks and wooden boxes. They **used** sticks as **tools** to pull in the bananas, clubs to get them from above, longer sticks to climb up and even **piled up** the **boxes** to climb up them.

Kohler concluded that chimps showed **insight behaviour**.

Imprinting involves Innate and Learned Behaviour

Most behaviour is a **combination** of innate and learned behaviour, e.g:

Baby ducks learn to recognise and follow the **first moving object** that they see during a **critical period** soon after they **hatch**. This is called **imprinting**. The first thing they see is usually their **mother**, but if they see a human, they can become imprinted on the human instead. Although they have an **innate instinct** to follow the first moving object they see, they have to **learn** what their 'mother' looks like, because they have no **innate experience** of what an adult duck looks like.

Practice Questions

Q1 Describe the difference between 'taxis' and 'kinesis'.

Q2 A cow touches an electric fence and gets a shock. From then on, it avoids the fence. What kind of conditioning is this?

Q3 Draw a table of differences between 'classical conditioning' and 'operant conditioning'.

Exam Questions

Q1 Describe a type of behaviour that appears to be a combination of genes and environment. [3 marks]

Q2 Suggest how operant conditioning could be used in dog training. [3 marks]

Oh bee-have...

It's a bit weird to think that a lot of our behaviour is instinctive and the rest of it we've probably been trained to do by our parents. I'm starting to wonder if we ever do anything just because we want to — after all, school is just one long system of rewards for some behaviours and punishments for others... might as well be a pigeon in a box, I reckon.

Courtship and Territory

In the animal kingdom, males court females to try and persuade the females to mate with them. They might do this by doing silly dances, bringing the female presents of food or even fighting other males. So not that different to humans, then.

Courtship involves *Showing Off*

Reproduction is vital for the survival of any species. In order to reproduce successfully, adults must choose a mate that is the **right species** and that is as **strong** as possible. This will give any young produced the greatest chance of **survival**. In most species it's the **male** that displays courtship behaviour to attract a female, although in some cases it's the other way round. To be chosen, the male must prove to a female that he's fit and virile, and this is where courtship behaviour comes in. Each species produces a **unique courtship signal**, so there's a huge **variety** of courtship behaviour seen in the animal kingdom. A courtship ritual usually includes one or more of the following:

1) **Sounds** — e.g. courtship **songs** in many bird species.
2) **Visual signals** — e.g. displaying colourful **plumage**.
3) **Behaviour** — e.g. performing courtship **dances**, bringing **gifts** of food. Some animals build **nests** or defend a **territory**.

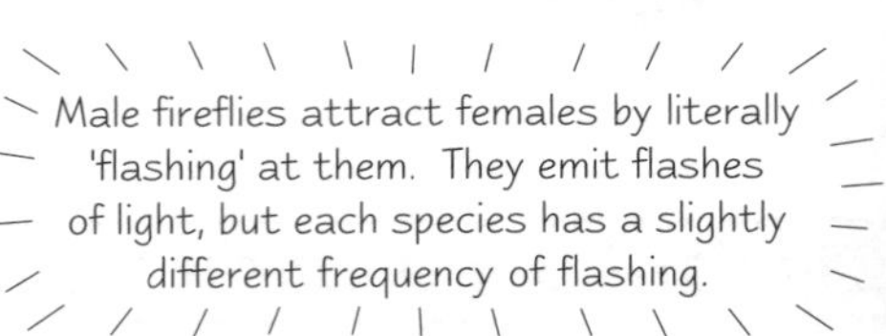

Male frigate birds inflate the sacs on their chests so that they look bigger and more impressive to females.

There are several reasons for the elaborate and prolonged courtship displays of some species:

1) The female needs to be sure that the male is of the **right species**. Mating between different species is a waste, as fertile offspring aren't produced. For this reason, even **closely related** species have **different** courtship rituals.
2) The male needs to be sure that the female is **sexually receptive** at the time. If she isn't, the male will know because she **won't respond** to the courtship behaviour.
3) The **quality** of the display may make a female more likely to choose a particular male. She wants to mate with a male who will pass on alleles to her young that will help them to **survive** — or that will help them to **breed** successfully in the future (see the section below on **sexual selection**).

Sign Stimuli are often Important in *Courtship Displays*

A **sign stimulus** is an **external signal** that triggers a particular response in another individual of the same species. The response is **innate** (genetically inherited), not learned, and occurs in **all** individuals of the species. For example, male **sticklebacks** attack other males, but court females. The **sign stimulus** triggering each behaviour is the abdomen of the other fish. If it's **red** (as in a male), the fish **attacks**. If it's **grey and swollen** (like a female with eggs) the male **courts**.

Experiments show that this happens even with models that don't look **anything** like a stickleback — it's only the colour and shape of the **abdomen** that's important.

This sort of behaviour often results from **innate releaser mechanisms**. This is a mechanism in the brain, which automatically triggers a specific behaviour or **fixed action pattern** in response to the sign stimulus.

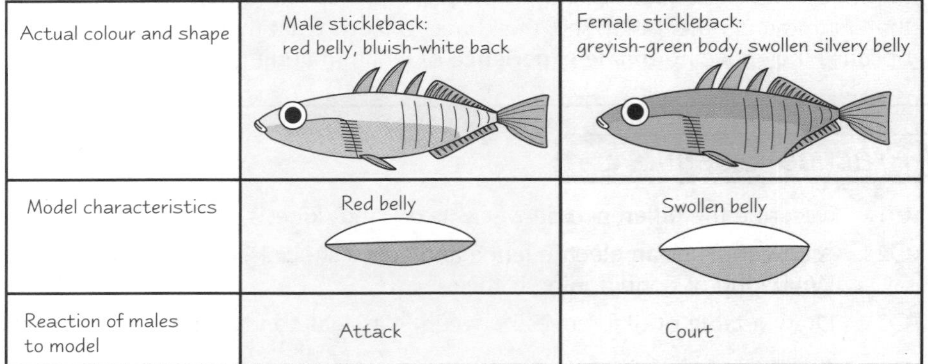

Actual colour and shape	Male stickleback: red belly, bluish-white back	Female stickleback: greyish-green body, swollen silvery belly
Model characteristics	Red belly	Swollen belly
Reaction of males to model	Attack	Court

Natural Selection can be *Sexual Selection*

Natural selection isn't just 'survival of the fittest', it's really '**breeding of the fittest**'. Some traits are seen in animals not because they help them to **survive**, but because they help them to **breed**. This is known as **sexual selection**. For example, **sexual selection** has resulted in male **peacocks** developing large and spectacular tail feathers because they **attract female peacocks**. A tail like the peacock's probably **reduces** its survival potential, as it's very visible to predators and slows the animal down. But it increases its **reproductive potential**, and so the trait has survived.

Courtship and Territory

It Always Helps if you Smell Nice

1) Courtship behaviour is influenced by a **complex interaction** of different **hormones**, with the **sex hormones** (not surprisingly) being particularly important.
2) Other hormones can also play a part in mating behaviour. For example, some species form long-term '**pair bonds**' — rarely, a male and a female may even stay together **for life**. One species that tends to form monogamous relationships for life is the **prairie vole**. Research on prairie voles has linked this behaviour with two **brain hormones**, **oxytocin** (in the female) and **vasopressin** (in the male).
3) Some animals secrete another type of chemical which helps them to attract a mate. These are called **pheromones**. A pheromone is defined as **a chemical substance produced by one animal that affects the behaviour of another**. Different types of pheromone have **different functions**, but one important type of pheromone is that produced by a female to alert males of her species in the area to the fact that she is **sexually receptive**. The pheromone then attracts the male and can also act as a **sign stimulus** to start mating behaviour. The advantage of pheromones is that they act over **long distances** and can attract males that the female isn't even aware of. They will also affect **all** the males in the area, allowing the female to select the 'best' male as her mate.

One of the best known pheromones is produced by the **female silk moth**. It can attract males from as far away as **10km** and the male can detect even a **single molecule** of it.

Territoriality is Having your Own Patch

Territoriality is a common behaviour in many different types of animal, including fish, birds and mammals. A **territory** is an area that is defended by an individual or group, which other members of the same species are **not permitted to enter**. Not all species have territories, but in those that do, territories have a **variety** of functions. They allow animals to **control access** to critical **resources**, such as nesting sites or food. And in many species, territories allow one sex (usually the **male**) to defend an area to which the other is attracted for **mating**.

In such species, having a desirable territory can be a **big advantage** in terms of breeding success. If a male can establish his own territory and keep other males out, it helps to convince the female that he's a **suitable mate**. A male that successfully defends a desirable territory is likely to be **strong** and **healthy**, and therefore an ideal mate. He will also have access to the **resources** needed to provide for any young produced. Males who have **not** been able to successfully find and defend a territory are **unlikely** to find a mate. Mating with a male that didn't have his own territory would be a **waste** for the female of such a species, as her young would be **unlikely to survive**.

Lions defend territories as a group, whereas leopards have their own individual territories.

Practice Questions

Q1 What is a sign stimulus?

Q2 Give an example of sexual selection.

Q3 What is a pheromone?

Exam Questions

Q1 State three advantages of courtship behaviour for the organisms involved. [3 marks]

Q2 (a) Explain the advantage to a female of producing pheromones. [2 marks]
(b) Suggest why it is usually the females of a species that produce the pheromones, rather than the males. [1 mark]

Q3 Suggest how territoriality may have evolved in a species by natural selection. [5 marks]

Hmmm, lady frigate birds certainly have strange taste...

In most species, the female chooses from several males who all want to breed with her. This is because it's usually the female that invests the most time and energy in the offspring. Males mate with as many females as possible, but the female only has one set of young at a time to pass on her genes. In species where only males care for young (e.g. the seahorse), they also choose the mates.

Hormonal Control of Reproduction

Sorry lads — these pages are pretty much devoted to the inner workings of the ladies.
Now one of them is bound to start going on about how easy you have it in comparison. Got a point, though...

The Human Menstrual Cycle is Controlled by Hormones

The human **menstrual cycle** lasts about **28 days**. It involves the development of a **follicle** in the ovary, the release of a **secondary oocyte** (immature ovum) and thickening of the **uterus lining** so a fertilised ovum can **implant**. If there's no fertilisation, this lining breaks down and leaves the body through the **vagina**. This is known as **menstruation**, and it marks the end of one cycle and the start of another.

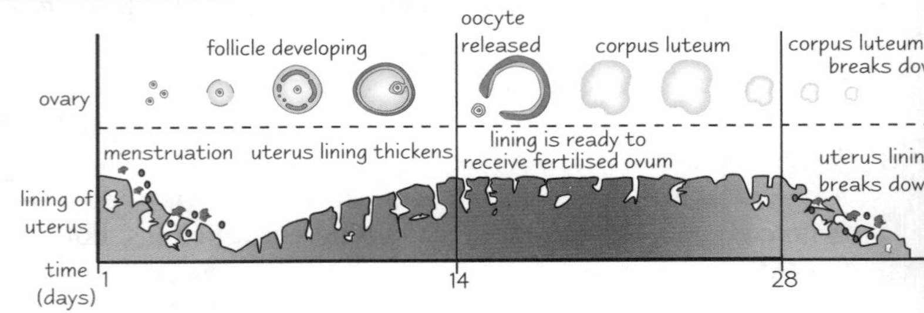

The menstrual cycle is controlled by four **hormones** — **follicle-stimulating hormone** (**FSH**), **luteinising hormone** (**LH**), **oestrogen** and **progesterone**. They're either produced by the **pituitary gland** or by the **ovaries**:

Hormones released by the anterior pituitary —

1) **FSH** is released into the bloodstream at the **start** of the cycle and is carried to the **ovaries**. It stimulates the development of one or more **follicles**, which in turn secrete **oestrogen**.
2) **LH** is released into the bloodstream around **day 12**. It causes **release** of the secondary oocyte (**ovulation**). When the oocyte bursts out, it leaves its **primary follicle** behind. LH helps the follicle turn into a **corpus luteum**, which is needed later.

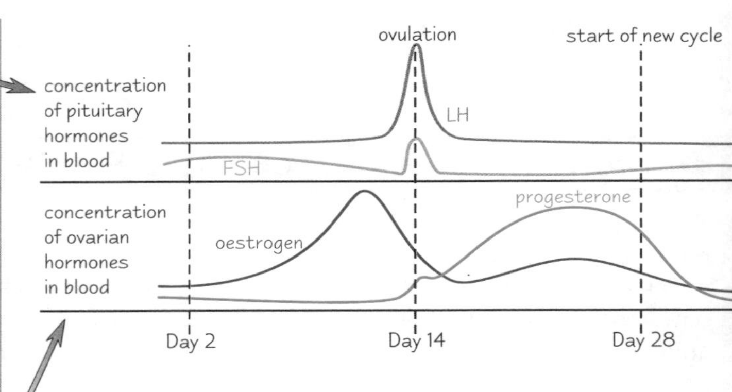

Hormones released by the ovaries —

1) **Oestrogen** is produced by the **developing follicle**. It causes the **lining** of the uterus to **thicken**. It also **inhibits** the release of **FSH**. This stops any **more** follicles maturing. But then a **peak** in oestrogen levels starts a **surge** in FSH and LH production, which triggers **ovulation**.
2) **Progesterone** is released by the **corpus luteum** after ovulation. Progesterone keeps the uterus lining **thick**, ready for implantation if fertilisation occurs. It also **inhibits** release of FSH and LH. If no embryo implants, the corpus luteum **dies**, so progesterone production stops and **FSH inhibition** stops. This means the cycle starts again, with development of a **new follicle**.

The secondary oocyte is released around day 14 of the menstrual cycle, and must be fertilised within 24 hours. If sexual intercourse leads to fertilisation, the fertilised ovum moves down the oviduct to the uterus where it implants in the wall. This takes up to 3 days.

Negative Feedback Loops regulate Hormone Concentrations

The stages of the menstrual cycle are carefully controlled by hormones, and **negative feedback loops** exist to keep the system in balance. In the menstrual cycle, **GnRH** (gonadotrophin releasing hormone, from the hypothalamus) stimulates the pituitary to release **LH** and **FSH**, which then stimulate release of **oestrogen** and **progesterone**.
But when levels of oestrogen and progesterone get **too high**, they **inhibit** the release of GnRH, so the release of **all the others** stops too. Then when levels **fall** again, the hypothalamus stops being inhibited and everything gets going again. This is what's meant by **negative feedback**.

Positive feedback loops also exist, e.g. uterus contractions that are caused by oxytocin stimulate more oxytocin to be released from the posterior pituitary during labour.

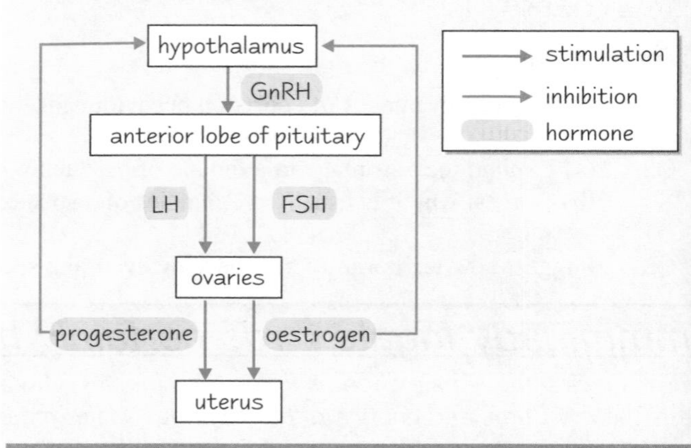

Hormonal Control of Reproduction

Hormones also control **Pregnancy**, **Birth** and **Lactation**

1) **Human chorionic gonadotrophin** (HCG) is secreted by the **embryo** and placenta. It maintains production of **oestrogen** and **progesterone** by causing the **corpus luteum** to develop. It also triggers the production of **testosterone** in a male embryo. Later during pregnancy (at 12-16 weeks), the **placenta** takes over from the corpus luteum in secreting progesterone and oestrogen and HCG levels begin to **fall**.

2) **Progesterone** levels increase during pregnancy, which helps to maintain the **uterus lining (endometrium)**. It also helps develop the uterus and breasts and **inhibits** contractions of the uterus. Levels then **fall** before birth.

3) **Oestrogen** levels rise during pregnancy too. Like progesterone, it helps maintain the **uterus** and **endometrium**. It also **inhibits FSH** so that no more oocytes are released during pregnancy. Oestrogen levels **carry on rising** at the end of the pregnancy and this overrides the effects of **progesterone**, allowing the uterus to start **contractions**.

4) **Oxytocin** is released by the **posterior pituitary gland** and stimulates **contractions** of the uterus before birth. Increasing oxytocin levels increase the **number** and **force** of the contractions and allow the baby to be born. Oxytocin also causes **milk** to be **released** when the baby is feeding.

5) **Prolactin** is secreted from the **anterior pituitary gland** and stimulates **milk production** (**lactation**) after birth.

The Hormone **HCG** is used for **Pregnancy Testing**

The hormone **human chorionic gonadotrophin** (**HCG**) is released by the **embryo** (and placenta) and excreted in the pregnant woman's **urine**. Pregnancy testing kits contain a strip which has coloured **HCG antibodies** attached to it, and this is dipped into the urine sample. If HCG is present in the urine, it **attaches** to the antibodies and a coloured line appears on the strip to show that the woman is pregnant. A **control line** also appears whether the woman is pregnant or not, to show that the test's worked.

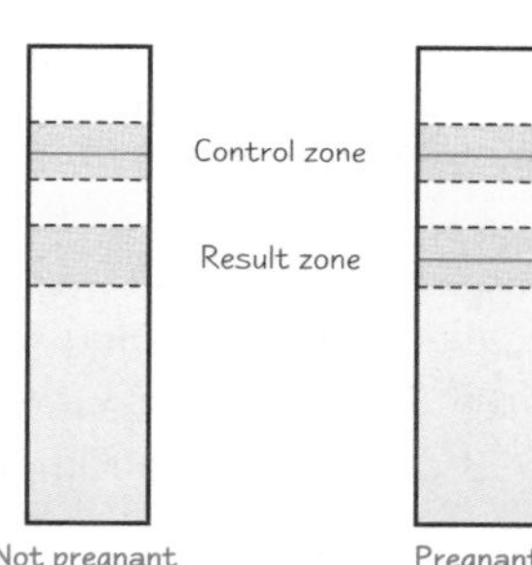

Practice Questions

Q1 Which hormones in the menstrual cycle are released by: a) the pituitary gland b) the ovaries?

Q2 Give an example of a negative feedback loop in the menstrual cycle.

Q3 Which hormone stimulates contraction of the uterus before birth?

Q4 Which hormone is used in pregnancy testing kits?

Exam Question

Q1 a) The graph on the right shows the levels of 4 hormones during pregnancy — HCG, oestrogen, prolactin and progesterone. Identify hormones A,B and C. [3 marks]

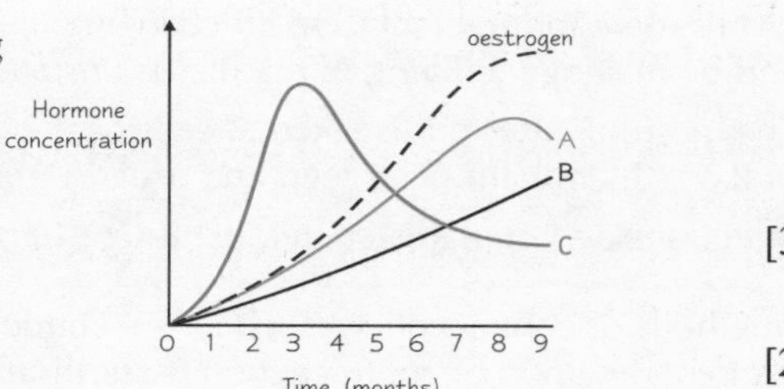

b) Where is the hormone prolactin produced and what is its main effect? [2 marks]

Sometimes it's hard to be a woman...

Aargh, so many hormones on this page! It's making me feel weepy and crave chocolate just reading about them all. The menstrual cycle is probably familiar to you from GCSEs, and there are four main hormones that control it — FSH, LH, oestrogen and progesterone. Then HCG, oxytocin and prolactin come into play during pregnancy and birth. Phew.

Conception, Contraception and Infertility

These two pages aren't too bad — the stuff on reproduction should be pretty familiar by now. I was going to say you do it every year, but then I realised that could be taken the wrong way...

The **Male Gametes** are Transferred to the **Female** during **Sexual Intercourse**

Sperm that are made in the testes collect in the **epididymis** and are stored in the **vas deferens**. In humans, **internal fertilisation** occurs (secondary oocytes and sperm fuse inside the female) so a process is required to deliver the sperm to an oocyte in an oviduct. This process is called **sexual intercourse** or copulation.

In the male the penis becomes **erect** through physical or psychological stimulation. **Nerve impulses** sent to the penis cause the arteries to **dilate**, filling the spongy **erectile tissue** with blood. During intercourse, **contraction** of the muscles around the **seminal vesicles**, **vas deferens**, **prostate gland** and **urethra** eventually result in the release of **semen**. This is known as **ejaculation**. The semen usually contains several **million** sperm in an **alkaline** fluid.

Sperm are usually released at the top end of the vagina near the **cervix** during intercourse. The alkaline semen protects the sperm from the **acidic conditions** in the vagina, and allows some of them to pass through the cervix into the **uterus**.

Capacitation Prepares the Sperm to **Fertilise** an Oocyte

In the first few hours after ejaculation, the sperm undergo a series of changes to get them ready to fertilise an oocyte. This process is called **capacitation**. It involves the removal of some of the **outer proteins** on the sperm, and the **reorganisation** of its plasma membrane. Capacitation makes the sperm more **mobile** and prepares its membranes for the **acrosome reaction** needed to enter the oocyte.

The sperm is now ready to **fertilise** an oocyte, as outlined below:

1) The sperm's **acrosome** releases enzymes to digest through the **follicle cells**.
2) **Membrane receptors** on the sperm then bind to receptors on the **zona pellucida**, causing more enzymes to be released from the acrosome.
3) The sperm **enters** the secondary oocyte, which then undergoes the **second division of meiosis** to give the **ovum** and a **second polar body**.
4) As soon as the sperm penetrates the cell surface membrane, **cortical granules** are released by the ovum. These cause the **zona pellucida** to form a barrier called the **fertilisation membrane**, to prevent any other sperm entering.
5) The sperm's nucleus **fuses** with that of the ovum to make a diploid **zygote** (23 chromosomes from the haploid sperm, 23 from the haploid ovum).

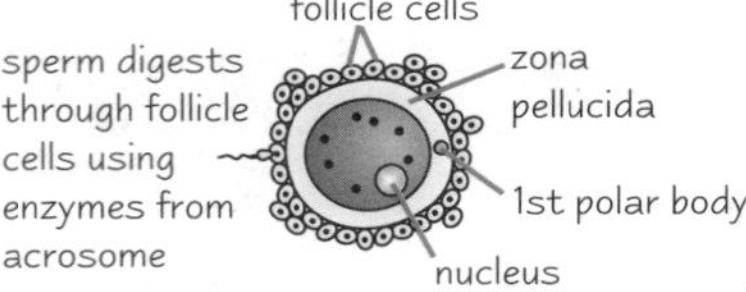

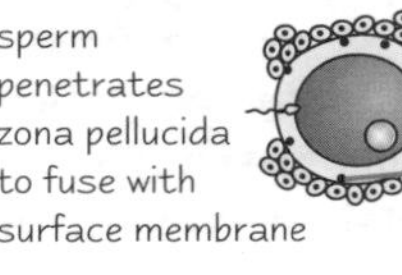

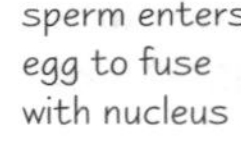

The Hormones **Oestrogen** and **Progesterone** are used for **Contraception**

Fertility can be **controlled** through the use of contraception.
Hormone-based **contraceptives** can be divided into the following categories:

1) The **combined pill** is taken orally and is a complex of synthetic hormones, usually **progesterone** and **oestrogen**. It mimics the hormone levels of **early pregnancy** and so inhibits the release of **FSH** (follicle stimulating hormone) and **LH** (luteinising hormone). This stops any **secondary oocytes** being released (ovulation).
2) The **mini pill** just contains **progesterone**. It's used by women who have a higher than normal risk of developing certain **health problems**, e.g. **thrombosis**, perhaps due to past problems or their family history. This is because the combined pill also slightly **increases** the risk. The mini pill doesn't always prevent **ovulation**, but it prevents **fertilisation** and **implantation** by changing the consistency of the **mucus** in the uterus to stop sperm swimming, and by making the **lining** of the uterus unsuitable for implantation.
3) The **morning-after pill** is taken after sex as an **emergency** measure (within a few days), e.g. if a condom has broken. It contains **oestrogen** and **progesterone**, which prevent **implantation**.
4) **Injections** and **implants** containing **progesterone** are also possible. They can stop **ovulation** for a few **months**.

Barrier methods of contraception (such as the **condom** and **diaphragm**) stop sperm from getting through to the cervix. Another method of contraception is **sterilisation**, where the **vas deferens** in men, or the **oviducts** in women, are cut and tied to prevent the release of sperm or ova.

Some methods of contraception are more **reliable** than others — for example, the mini pill only works if it's taken at the same time every day. (Sterilisation is the only **permanent** method.) Some methods also have added **advantages** — for example, barrier contraceptives stop you getting **sexually transmitted infections**. The combined pill can reduce **PMS** and **acne**, but it has also been linked to **thrombosis** and some types of **cancer**. The choice is yours...

Conception, Contraception and Infertility

Sometimes **Conception** can't happen without **Help**

Infertility can be caused by problems in **either** the male **or** the female **reproductive system**.
This section deals with **female** infertility, which is most commonly caused by the following conditions:

1) **Abnormal ovulation**, or irregular release of a secondary oocyte from the ovary. Normally one oocyte is released **each month** under the direction of several **hormones**. But if any of these hormones aren't functioning properly, ovulation will happen **irregularly** or **not at all**.
2) A **blockage** in one or both of the woman's **fallopian tubes**. This means that the oocytes can't get to the uterus and sperm can't reach the oocytes.

Abnormal Ovulation is usually Treated using **Drugs**

The development of oocytes and their release in ovulation is controlled by hormones called **gonadotrophins**, which are produced in the **pituitary gland**. If these hormones are at low levels, oocytes might not be released. Treatment for this condition may involve the injection of extracted or synthetic **gonadotrophins**, or taking **drugs** which stimulate their production (e.g. **clomiphene**).

Blocked Fallopian Tubes Require **In Vitro Fertilisation**

If a **blockage** is stopping the oocyte and sperm meeting, or the oocyte getting to the uterus, then **in vitro fertilisation** (**IVF**) often gives the best chance of conceiving a child. This is where sperm and oocytes are taken from the parents (or possibly from a donor) and fertilisation happens in a **laboratory** instead of inside the woman. The embryos can then be **implanted** back into her uterus once fertilisation has happened. The full procedure is shown on the right.

This is a **complex** and **expensive** process and there's **no guarantee** of success. The technique involves the use of **drugs** and minor **surgical procedures**, so it carries more **risk** for the mums and babies than a natural pregnancy does. Because it's quite likely that **none** of the embryos will develop, until recently **several** embryos were implanted back into the uterus to increase the chances of at least one developing into a baby. But the high number of **multiple births** (twins and triplets) associated with IVF has meant that **fewer embryos** are now implanted back into the uterus. Multiple births are **more risky** for both the mum and the babies.

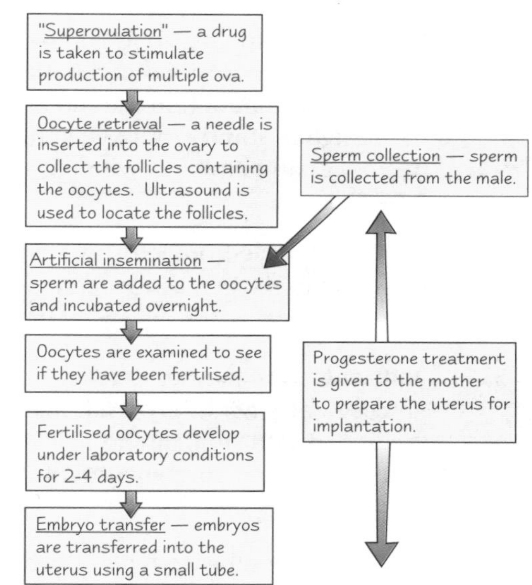

Practice Questions

Q1 What is the role of the sperm's acrosome in fertilisation?

Q2 Name two types of oral contraceptive.

Q3 What is in vitro fertilisation (IVF)?

Exam Questions

Q1 Describe two treatments available for the correction of abnormal ovulation, and explain how they work. [4 marks]

Q2 Describe the stages in IVF treatment. [7 marks]

Let's talk about delivery of sperm to an oocyte in an oviduct, bay-bee...

OK it's not as catchy, but it's more scientifically accurate and that's what counts in this book. What a hassle sex is — people who don't want babies getting pregnant, people who do want babies not getting pregnant. Plus you can get all kinds of nasty diseases. Personally when I start getting broody I'm just going to grow a baby at the end of my arm like a strawberry plant.

Pregnancy

You've read all about how not to get pregnant, and what you could try if you want to but can't get pregnant. Now here's a lovely couple of pages about pregnancy itself. Although mostly it seems to be about the placenta. Yuck.

The **Placenta** allows Transfer of Materials between Mother and Fetus

The **placenta** starts to form shortly after the **blastocyst** (the mass of cells formed from the zygote) implants in the uterus wall. The blastocyst develops into an **embryo** and starts to form a layer called the **chorion** which grows into the **endometrium** (lining) of the uterus. Tiny projections called **chorionic villi** are produced with capillaries inside containing the fetus's blood. Spaces develop around the villi in the endometrium called **sinuses**, which become filled with the mother's blood. The villi themselves are filled with **mitochondria** and lined with **microvilli** to aid transport of substances between the mother and the embryo, which soon develops into a **fetus**. The two blood supplies mustn't mix as there are differences in blood **pressure**, and possibly blood **groups**, but they flow very **close** to each other to allow easy transport of materials.

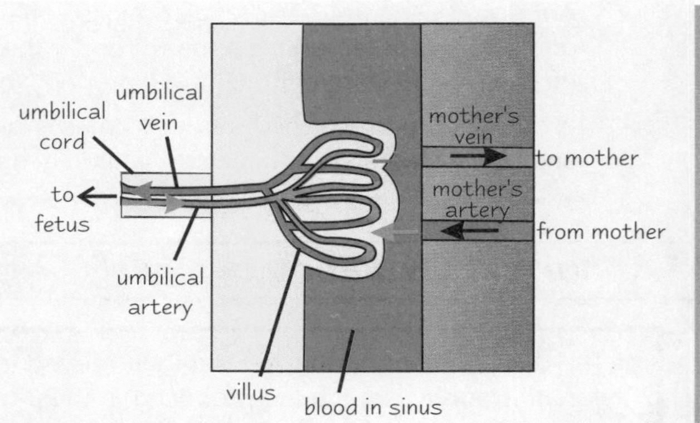

The **Structure** of the Placenta helps its role in **Exchange of Materials**

The placenta is involved in the **transfer** of the following substances:

1) **Oxygen** from mother to foetus — **oxygenated** blood enters the **sinuses** from the mother's arteries and flows next to the fetus's blood in the **placental capillaries**. It diffuses along a **concentration gradient**, from the higher concentration in the mother's blood to the lower concentration in the fetus's blood. Fetal **haemoglobin** has a **higher affinity** for oxygen than the haemoglobin in the mother's blood, which means that fetal haemoglobin combines with oxygen at the same partial pressure that maternal haemoglobin releases it.

2) **Carbon dioxide** from fetus to mother — carbon dioxide **diffuses** out of the fetal capillaries in the opposite direction to oxygen as the fetal blood flows past the mother's blood.

3) **Nutrients** from mother to fetus — the developing fetus needs the same nutrients as the mother, including **glucose**, **fatty acids**, **glycerol**, **amino acids**, **vitamins** and **minerals**. Glucose is passed to the fetus by **facilitated diffusion**, amino acids are transferred by **active transport**, and vitamins and minerals move across to the fetus by a **combination** of diffusion and active transport.

4) **Water** from mother to fetus — water passes into fetal blood by **osmosis**.

5) **Urea** from fetus to mother — this waste material **diffuses** into the mother's blood to be excreted.

6) **Antibodies** from mother to fetus — some antibodies can pass across the placenta from mother to fetus, giving the fetus **temporary passive immunity** to some diseases when it's born.

As well as being a **transport organ**, the placenta also acts as an **endocrine gland** during pregnancy. It secretes **oestrogen**, **progesterone** and **HCG**, which amongst other things help to maintain the **endometrium** and develop the uterus and breasts.
The placenta also acts as a **barrier** against harmful bacteria and most viruses. It can't stop the **rubella virus** or **HIV** though, and it's not a barrier to harmful substances and drugs such as **alcohol**, **nicotine** and **heroin** which can harm the fetus.

Transport method	Example
diffusion	O_2, CO_2, urea
facilitated diffusion	glucose
active transport	amino acids
osmosis	water

Pregnancy

A Woman's body **Functions Differently** during **Pregnancy**

While a baby is developing inside the mother's uterus, the **mother's body** has to **adjust** to the demands of pregnancy. Some of the main changes are:

1) The mother puts on **weight**. This might seem obvious, but it's **not** just the weight of the baby. The **uterus wall** and the **breasts** increase in size. The **placenta** adds some extra weight. There will be an increase in the amount of **blood** and **body fluids**, and **fat** will be stored up ready to provide extra energy for **breast feeding**. **How much** weight a woman puts on depends on her **height** and **starting weight**. This is calculated as her **BMI** (**Body Mass Index**). The graph shows **recommended weight gain** for different BMIs.

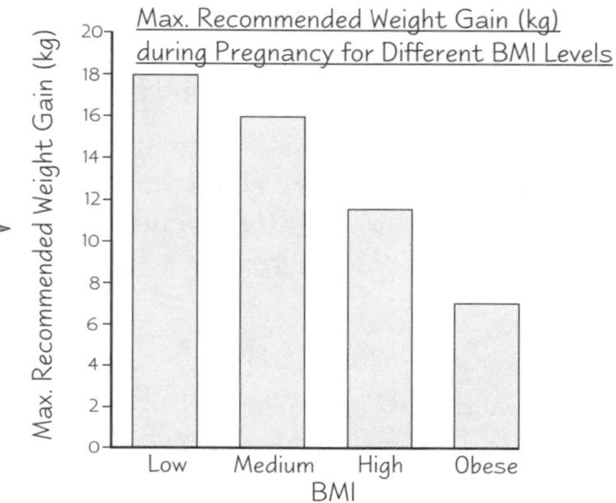

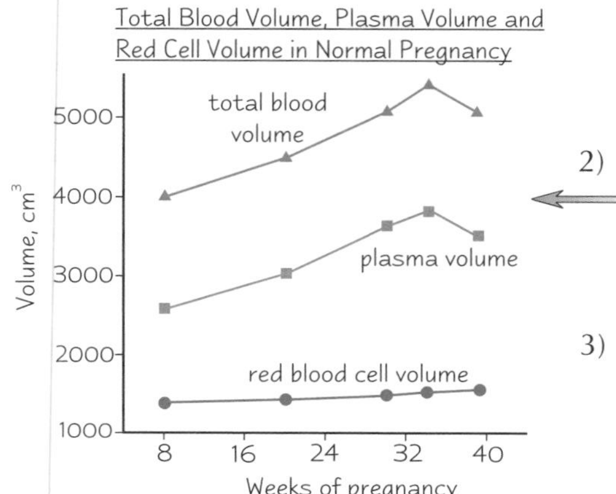

2) **Plasma volume** and **red blood cell mass** increase. This is to meet the **metabolic demands** of pregnancy and breast feeding. The effects are shown in the graph.

3) **Cardiac output** increases. The mother's **heart rate** goes up and so does the **stroke volume** (how much blood is pumped out with each beat). Again, this helps to meet the extra **energy demands** on the mother's body.

4) **Kidney function** alters. The mother's body has to deal with the **baby's wastes** in addition to her own. To cope with this, her kidneys filter fluid at an **increased rate**.

Pregnancy puts quite a strain on a woman's body, especially the **heart** and the **kidneys**. A healthy woman will be able to deal with these extra demands, but this is one reason why **regular medical checks** during pregnancy are essential.

Practice Questions

Q1 Why is it important that a pregnant woman's blood doesn't actually mix with that of her fetus?

Q2 Name two viruses that can pass across the placenta from mother to fetus.

Q3 What happens to the plasma volume of a pregnant woman?

Q4 Why does the mother's cardiac output increase during pregnancy?

Q5 Why does the mother's kidney function increase during pregnancy?

Exam Questions

Q1 a) List the substances that pass from the mother to the fetus and vice versa via the placenta. Explain how each substance moves from one bloodstream to the other. [12 marks]

b) Give two other functions of the placenta, besides transport. [2 marks]

Q2 Explain why it is normal for a pregnant woman to put on weight during pregnancy. [6 marks]

So whether we remember it or not, we've all tasted our own wee...

The placenta's important. It lets the blood of mum and fetus get close together without mixing, so that stuff can pass across. There are six main things that are exchanged across it (try writing down all six now without looking — people always seem to forget one). Plus it secretes some very important hormones and acts as a barrier to stop harmful stuff crossing.

Growth and Development

There are a few big words on these pages. I know it all sounds confusing, but there's nothing here you haven't covered before. And puberty? Well, pah — been there, done that, can point to the body parts that did interesting things.

Thyroxine secretion is controlled by the Hypothalamus and Pituitary Gland

Hormones help control our **growth** and **development**. The hormones involved are produced by the **hypothalamus** and the **pituitary gland**, which are located at the base of the brain.

The hypothalamus is a regulatory centre that secretes **thyrotrophin releasing hormone** (**TRH**). TRH affects the **anterior lobe** of the **pituitary gland**, causing it to release **thyroid stimulating hormone** (**TSH**) into the bloodstream. TSH in turn affects a gland near the trachea in the neck called the **thyroid gland**. It stimulates the thyroid gland to release another hormone called **thyroxine**, which increases **metabolism**.

The whole system has a **negative feedback loop** so that levels of these hormones are kept in check and growth is regulated:

Thyroxine **inhibits** the production of **TRH** and **TSH**, so if the metabolic rate gets too **high** due to too much thyroxine, less TRH and TSH are released. Thyroxine levels then **fall** and so does the **metabolic rate**. If the metabolic rate falls **too much**, the **low** thyroxine concentration means that **more** TRH and TSH are released. This **increases** the thyroxine levels again.

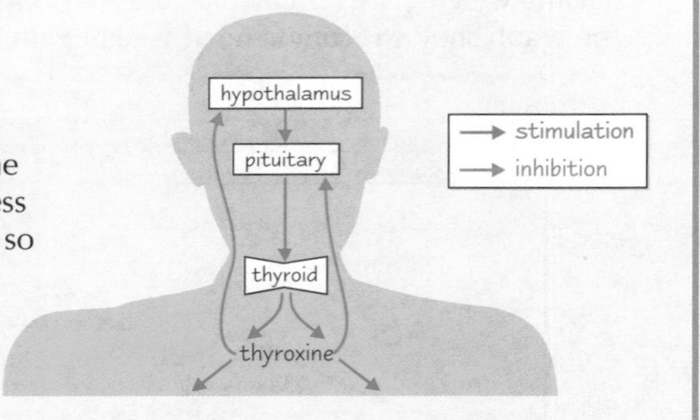

Thyroxine controls Growth and Development

1) The **thyroid gland** consists of a network of **blood capillaries** with **follicles** between them. The **follicle walls** are a single layer of cells that secrete **thyroglobulin** into the **lumen** at the centre of the follicle, where it's **stored**.
2) Thyroglobulin is a large protein-based molecule that contains **iodine**. It's taken up by the cells when **thyroxine** is being made. Enzymes convert the thyroglobulin into the smaller molecule thyroxine, which can enter the **blood capillaries**.
3) Thyroxine affects nearly **all** the cells in the body by increasing their **metabolic rate**. It increases the rate of **cellular respiration**, and the rate of protein, carbohydrate and fat **metabolism**. This leads to **growth**, and also to the release of more **heat** and **energy** by the cells.

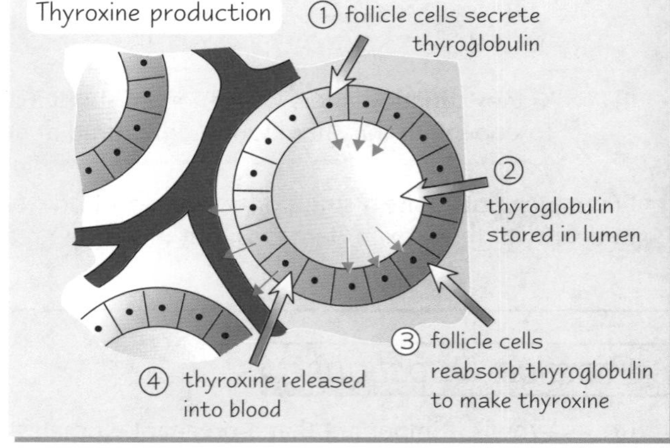

4) Thyroxine is particularly important in the growth of **teeth** and **bones** in children, as well as in **brain development**. Women who don't get enough iodine when they're pregnant can't make enough thyroxine, and can have babies whose **nervous** and **skeletal systems** aren't properly developed. Adults suffering from a lack of iodine in the diet develop a **goitre** (enlarged thyroid gland). Adults with an underactive thyroid lack energy, become depressed and may gain weight.

Growth throughout the body is also promoted by a non-specific hormone simply called **growth hormone** (**GH**). It's produced by the **anterior lobe** of the **pituitary gland**, and its release is controlled by the **hypothalamus**. It stimulates **protein production**, and it's important that the correct amounts of GH are produced if a normal **body size** is to be reached.

The development of **secondary sexual characteristics** is caused by **testosterone** and **oestrogen** produced by the **testes** and **ovaries** respectively. See the next page for more on how these characteristics are controlled by hormones during puberty.

Growth and Development

Absolute and **Relative Growth Rates** are measures of growth

1) **Absolute growth rate** is the amount of growth **in a given period of time**. A curve showing the **total** amount of **growth** would look different to one showing absolute growth rate. It would increase steadily, because organisms don't tend to shrink, and it'd probably be sigmoid in shape. But a **curve of absolute growth rate** can increase and decrease, because an organism can grow more quickly for a while and then grow more slowly.
2) **Relative growth rate** relates the **absolute growth rate** to an organism's **size**, e.g. if a 10g and a 20g organism both grow 1g per day, the 10g organism has **twice** the relative growth rate. Relative growth rate graphs can show a **decrease** over time.

In humans, different growth curves are produced depending on what's measured — e.g. absolute mass or height over time, relative growth rates of certain parts (e.g. the brain or reproductive organs), etc.

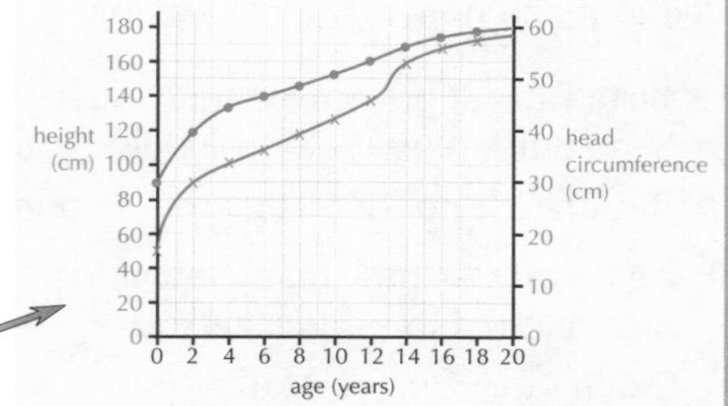

Puberty is the onset of **Sexual Maturity**

Puberty is the stage between childhood and adulthood. It's started by the release of **GnRH** (gonadotrophin releasing hormone) from the **hypothalamus**, which stimulates the **pituitary gland** to release other hormones:

1) In females **FSH** and **LH** are released, which causes the ovaries to release **oestrogen** and start the **menstrual cycle**.
2) In males **FSH** and **LH** are released, which causes the testes to release **testosterone** and start to produce **sperm**.

These **sex hormones** cause the development of the **sexual characteristics**:

Female	Male
Menstruation	Penis and testes grow
Breasts grow	Muscles develop
Pubic hair	Pubic and body hair
Hips widen	Larynx enlarges so voice deepens

A **long** pre-puberty stage in humans means there's a long time for the **brain** to develop before maturity. There is also a long time for **learning** and developing **complex behaviour**. This has been important in **evolutionary terms**, as humans have been able to use their intelligence to survive.

Ageing is associated with less **Efficient Functioning** of the Body

The effects of **ageing** are sadly inevitable, and include:

1) Changes in **physiological functions** — organs begin to function **less efficiently**. For example, there's a decline in **cardio-vascular** efficiency and capacity with old age.
2) **Tissue degeneration** — structural proteins become **harder** and **less elastic**.
3) Accuracy of **DNA replications** declines, so some **genetically abnormal** cells start to be formed during **mitosis**. These cells won't be able to do their jobs properly. They might even die or become **cancerous**.
4) Decline in the effectiveness of the **immune system** — making many diseases **more serious** in later life.

Practice Questions

Q1 Where in the body are the thyroid, hypothalamus and pituitary gland?

Q2 Explain the difference between absolute growth rate and relative growth rate.

Q3 Why do humans have such a long pre-puberty stage?

Exam Questions

Q1 From the graph shown on the right, find:
a) the age (in years) at which rate of growth is highest. [1 mark]
b) the rate of growth at 4 years of age. [1 mark]
c) when growth has stopped. [1 mark]

rate of growth (cm / year) — age (years)

Q2 Many people become less mobile and experience more health problems in their old age. Briefly explain why, in terms of changes within the body. [8 marks]

I always thought puberty was associated with **im**maturity...

Actually, that kind of comment isn't going to wind you lot up any more, is it? Most of you are safely past all that, and sighing in disgust at the teenaged antics of the fifteen year olds in the corridor. You're probably nodding smugly in agreement, but don't get too pleased with yourselves — just round the corner waits ageing, and that's even less fun.

Population Size and Structure

Pages 80-81 dealt with the way in which a population grows in a laboratory culture. In the real world, however, things are a little more complicated. Especially when we're talking about more complex organisms than bacteria...

Population Change involves lots of **'Rates'**

If you want to **predict** what will happen to a population, you need to know how many individuals are being **added** to it and how many are **leaving** it. Changes in population size are caused by the following factors:

1) **Birth rate** — how many individuals are being **born** (per unit of population) in a **given time**. Sometimes referred to as **natality rate** (which is probably better, because plants, for example, aren't 'born').
   Birth rate = **no. of births / no. in population**
2) **Death rate** — how many individuals are **dying** (per unit of population) over a **given time**. Also referred to as **mortality rate**.
   Death rate = **no. of deaths / no. in population**
3) **Immigration** — the number of individuals **joining** the population from somewhere else in a **given time**.
4) **Emigration** — the number of individuals that are **leaving** the population to move elsewhere in a **given time**.

Population growth = (no. individuals born + no. of immigrants) – (no. of individuals dying + no. of emigrants)

Birth rate and **death rate** can be used to predict population growth as they usually remain **fairly constant**, but it's hard to predict **immigration** to and **emigration** from a population without **long-term data**. This kind of data can be used to plot **population growth curves**, which you need to be able to **interpret**. A typical population growth curve is shown on the right.

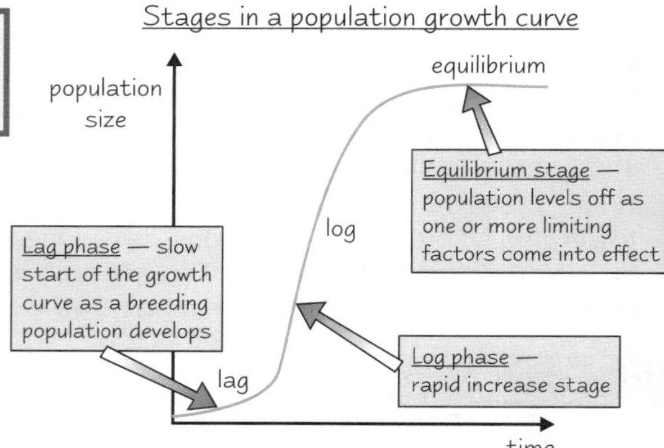

Population Data can be used to plot **Survival Curves** and **Age Pyramids**

If data are available about **death rates** at **different ages** in a population, a **survival curve** can be plotted. This shows how the death rate in a population changes with **time**, i.e. whether the death rate increases, decreases or stays the same as the population gets **older**. Data like this can also be used to find the **life expectancy** of the population — this is the average age at which individuals in the population die.

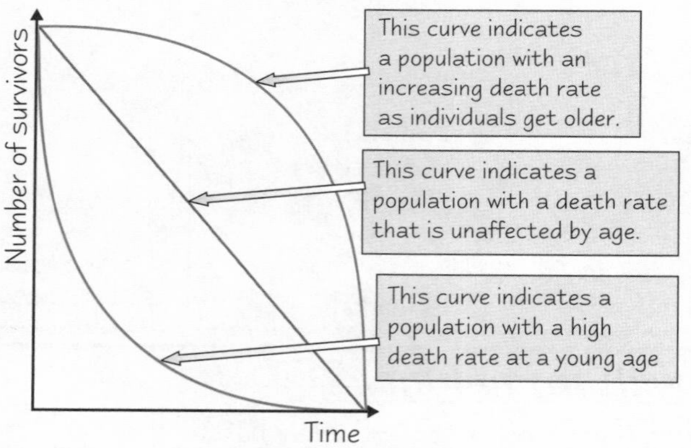

Useful information is also provided by plotting an **age pyramid** of the population. The height of the pyramid represents all the ages present in the population, with the youngest at the bottom and the oldest at the top. The **width** of the pyramid represents the **number of individuals of that age** in the population. The shape of the pyramid indicates **future trends** in the population.

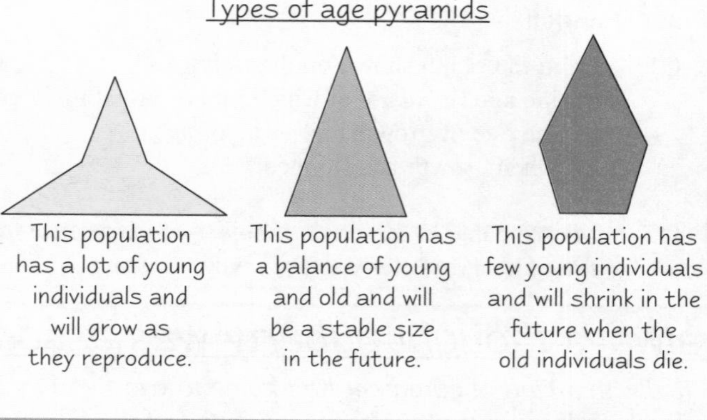

Population Size and Structure

The Demographic Transition Model shows Trends in Human Populations

The **demographic transition model** of human populations indicates trends in **birth** and **death rates** over a **long period** of time, as the population becomes more **developed**. It consists of **four** stages:

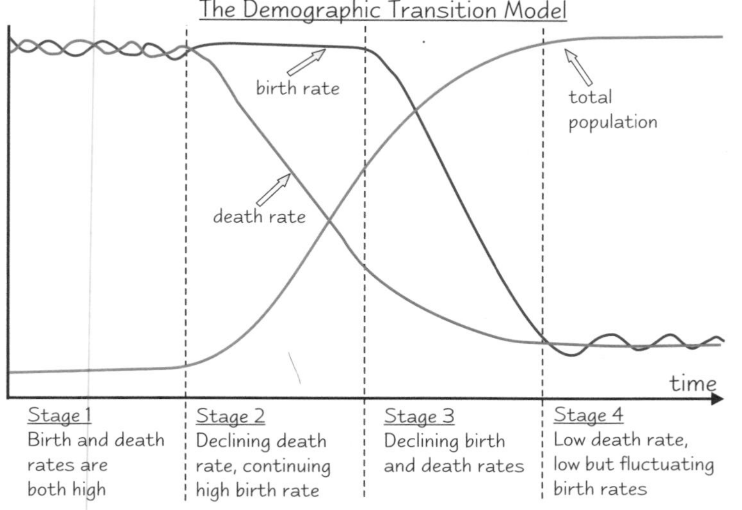

Stage 1 is when the level of development is **low**. There are **high** birth and death rates. The **high birth rate** is related to the absence of **family planning**, high **infant mortality**, and the need for lots of children to earn money for the family and to take care of their parents when they're old. The **high death rate** is due to **poor sanitation** and lack of **clean water** (which both cause **disease**) and shortage of **food**. The population is **stable**, as birth and death rates **balance**.

Stage 2 occurs when sanitation, water supply, food supply and medical care **improve**. There's still a **high birth rate**, but the **death rate** drops so the population **grows rapidly**.

Stage 3 is linked with an increase in **family planning** and the fact that, with **low infant mortality**, there's no need to have lots of children. The **birth rate falls** and, with the death rate remaining **low**, the population growth rate **slows**.

In **stage 4** the birth rate and death rate are both **low**, and the population becomes **stable** again.

Practice Questions

Q1 State the four factors that influence the size of a population over time.

Q2 What is meant by birth rate?

Q3 What is shown by an age pyramid?

Exam Questions

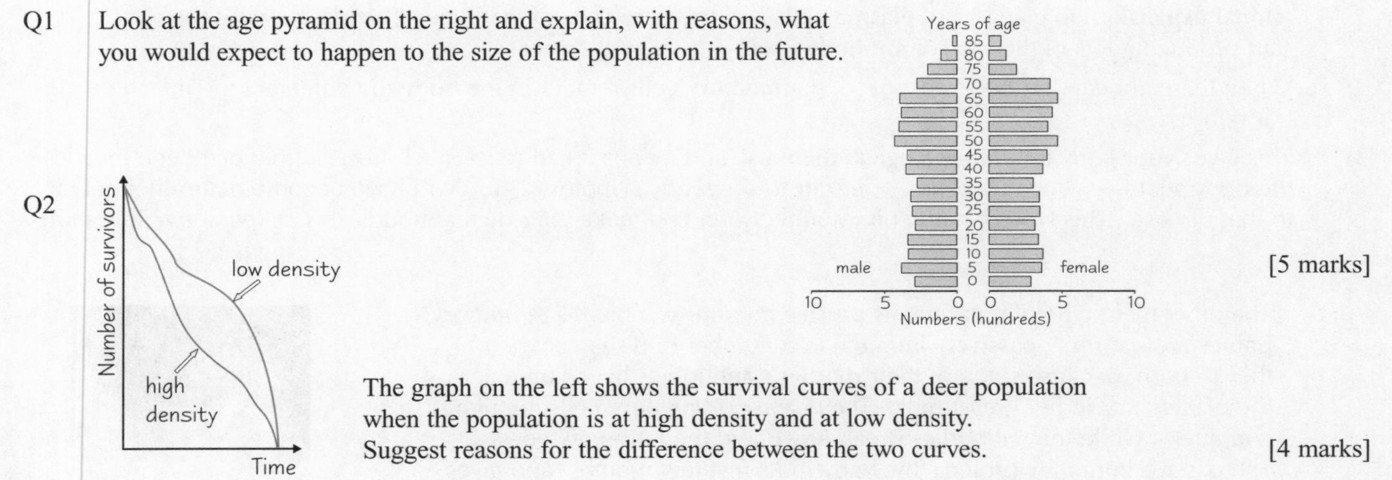

Q1 Look at the age pyramid on the right and explain, with reasons, what you would expect to happen to the size of the population in the future. [5 marks]

Q2 The graph on the left shows the survival curves of a deer population when the population is at high density and at low density. Suggest reasons for the difference between the two curves. [4 marks]

Just the thing to cheer you up — lots of graphs about death...

... no wonder you kids enjoy doing biology revision so much. You've got to learn the basic shapes of these curves and pyramids, but understand that they won't always look the same — it depends what's going on in the individual populations. You need to be able to look at a curve or pyramid and use it to say what's happening in the population it represents.

Infectious Disease and Immunity

Well isn't this a nice varied section — from courtship through pregnancy and population growth and on to infectious disease, all within a few pages. What on earth will be next — boomerang trajectories? Cutlery maintenance? Who knows.

Coughs and Sneezes Spread Diseases...

Any microorganism that causes disease is called a **pathogen**. For a pathogen to affect you, it has to actually get **inside your body** (or at least into your skin). Your body has a number of entry points, and for every one there are some microbes that use it to get in.

- **Droplet infection** occurs when you cough or sneeze out droplets containing pathogens, which then get breathed in by other people. **Chickenpox**, **influenza** and **TB** are examples of diseases spread in this way.
- Microbes can also get transferred by **contact**. This can be **direct** (one person touching another), or **indirect** (a person touching something that's been touched by another person). Examples include **conjunctivitis, athlete's foot** and **cold sores**. Transmission is more likely if the skin is broken, as it's a lot easier for the pathogen to get in.
- Disease can also be passed on via **food** and **drink**. **Food poisoning** is the obvious example, and **cholera** is another.
- Other ways that pathogens can enter the body are **sexual transmission** (e.g. HIV) and via **carriers (vectors)** — usually insects that bite humans (e.g. **malaria** is spread by *Anopheles* mosquitoes).

You will be Naturally Immune to some Diseases

To understand **immunity**, you need to know about **antibodies** and **antigens**. Antibodies are chemicals that **destroy microbes**, and they're produced by a type of **white blood cell** called a **B-lymphocyte**.
The antibodies 'recognise' the disease microbe by the pattern of chemicals on the pathogen's surface. These chemicals are called **antigens**. There are many different types of B-lymphocytes, each producing antibodies against a **different antigen**.

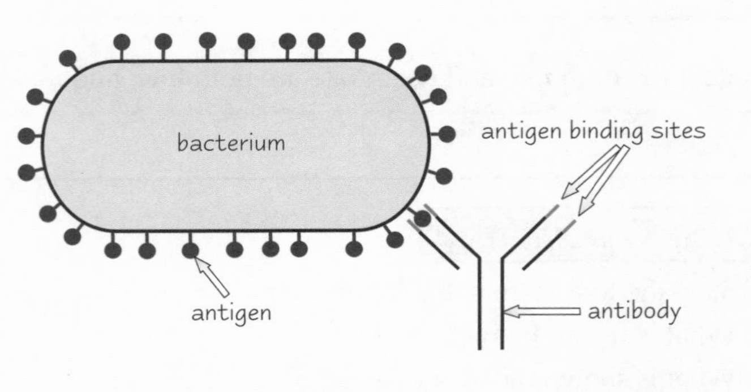

1) If a B-lymphocyte recognises a pathogen of 'its type', the cell starts **dividing rapidly** by mitosis (**clonal expansion**) to give lots of **plasma cells** and **memory cells** of that type. The plasma cells then start producing lots of the type of **antibody** needed to fight the disease.
2) When the pathogen has been destroyed, the **memory cells** remain in the body to counteract the infection if it returns.
3) This gives your body an **immunological memory**, and means it can react much **faster** if that pathogen invades the body again — probably before you start to show any symptoms. You will have become **naturally immune** to that disease. This type of natural immunity, when you make your own antibodies, is called **active immunity**.

The other main type of immunity is **passive immunity**, which is **temporary**. Babies are **naturally passively immune** to a number of diseases when they're born, because some of their **mother's** antibodies have come across the placenta or in her breast milk. They have no '**memory cells**' though, and after a while the antibodies **break down** and the immunity goes. The passive immunity protects the baby in its first few months, and gives it a chance to build up its own **active immunity** to common pathogens.

Billy would have been pleased about his passive immunity, if only he could breathe.

Infectious Disease and Immunity

Pathogens can Outwit the Immune System

There are some diseases that you **can't** become permanently immune to, either naturally or after **vaccination** (see next page). The **common cold** and **flu** are examples. The microorganisms avoid the immune response because they constantly evolve into new strains with **different antigens**. When the new strains get into the body, the B-lymphocytes can no longer recognise them and the build up of antibodies has to start all over again.

Vaccines help out our Immune Systems

While your B-lymphocytes are busy dividing to build up their numbers to deal with a pathogen, you suffer from the disease. **Vaccination** can help avoid this. It lets your body increase its level of the B-lymphocyte and antibody needed to fight a particular disease, **without** the microbe being present and increasing in numbers at the same time. This means you get the **immunity** without getting any **symptoms**.

Vaccines give you **active artificial immunity**. It's **active** immunity because your body makes its own antibodies, and will produce memory cells to help prevent the disease in the future. And it's called **artificial** immunity because you haven't actually **caught** the disease in the natural way — your body has been 'tricked' into reacting against a harmless pathogen.

Sometimes people need to be given **'instant immunity'** (vaccination takes a while to be effective). For example, if you get a bad cut you're often immunised against **tetanus**. This involves injecting the patient with the **tetanus antibodies** themselves, rather than with a harmless pathogen. It gives you **artificial** immunity **immediately**, but it's also **passive** immunity because your B-lymphocytes haven't been involved in producing the antibodies and there will be no memory cells. The immunity will wear off when the antibodies break down.

Vaccines not only protect the people vaccinated, but, because they reduce the **occurrence** of the disease, those **not** vaccinated are also less likely to catch it. This is called the **herd immunity effect**.

Practice Questions

Q1 What is a pathogen?

Q2 What type of blood cell produces antibodies?

Q3 What is meant by the 'herd immunity effect'?

Exam Question

Q1 The graph on the right shows the number of cases of polio reported each year since 1991 in the Indian province of Tamil Nadu. A vaccination programme was in place throughout this period.

a) The same vaccine programme was in place over the years 1991-1993, yet the cases of polio dropped dramatically. Suggest a reason for this. [1 mark]

b) There are now very few cases of polio in Tamil Nadu. Explain why the vaccination programme should **not** be discontinued. [1 mark]

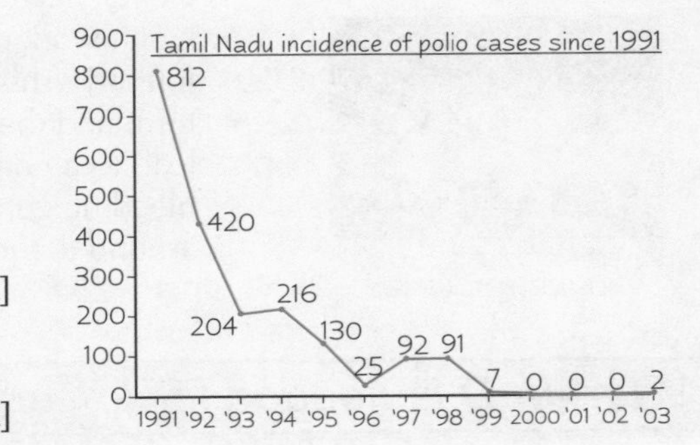

Immunity — another good idea from God / random mutations / aliens...

Delete as appropriate, according to your own belief system. I'm a big believer in political correctness, me. Personally I think it was the aliens, but I'm not saying that's right, it's just my opinion. It's also my opinion that you need to get this page learnt, as thoroughly and as quickly as you can. See, some of my opinions are pretty sensible. Others — not so much.

Balanced Diet

Food is made up of a mixture of carbohydrates, lipids, proteins, vitamins and minerals. They're all really important for being healthy. If you don't get enough of them you could be in trouble... bleeding gums and bow legs here we come...

Food is a Complex Mixture of Organic and Inorganic Compounds

Organic food compounds belong to **four main groups**:

1) **Carbohydrates** include **sugars** and **polysaccharides** (e.g. starch).
2) **Lipids** include fats and oils, and the **fatty acids** they're made of.
3) **Proteins** and the **amino acids** they're made up of.
4) **Vitamins**.

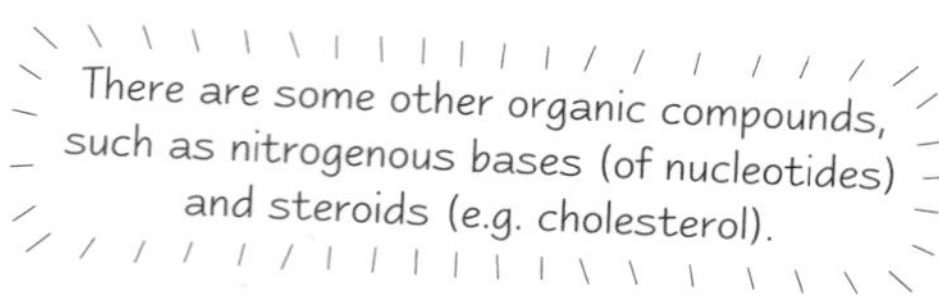

And there are two types of inorganic compound:

5) **Mineral salts** contain useful **mineral ions**.
6) **Water**.

A **balanced diet** has all these different compounds in the right proportions for good health. Of course, these proportions are different for each person — they change because of things like age, gender and lifestyle. Some compounds can be made by the cells in the body, but you've got to get the others from your diet — these are called **essential nutrients** and they include vitamins, minerals and some types of amino acids and fatty acids.

Vitamin A is needed for healthy Lining Tissue and Eyes

Vitamin A (retinol) is a fat-soluble vitamin found in dairy products and fish liver oils. You can make vitamin A from a group of compounds called the **carotenoids**, found in orange and yellow coloured fruit and vegetables (e.g. carrots).

Vitamin A is needed to keep lining tissue (epithelium) healthy. It's also used to make the visual pigment called **rhodopsin**, found in the rods of the retina (see pages 22-23). Vitamin A deficiency leads to a kind of blindness linked with an eye condition called **xerophthalmia**. The skin also becomes hard and flaky, making it prone to infection. Nasty.

Vitamin C is needed to maintain healthy Connective Tissue

Vitamin C (ascorbic acid) is a water-soluble vitamin especially found in citrus fruits, green vegetables and potatoes. Vitamin C is needed for the enzyme **hydroxylase** to work properly.

Hydroxylase is essential in maintaining the **collagen** fibres in **connective tissue**. It converts an amino acid called **proline** into another one called **hydroxyproline**. **Hydrogen bonds** hold the polypeptide chains of the collagen fibres together. These bonds can only form from one hydroxyproline to another hydroxyproline on a different fibre.

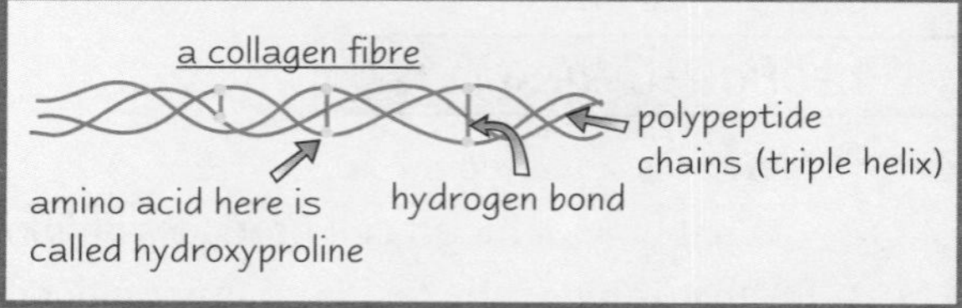

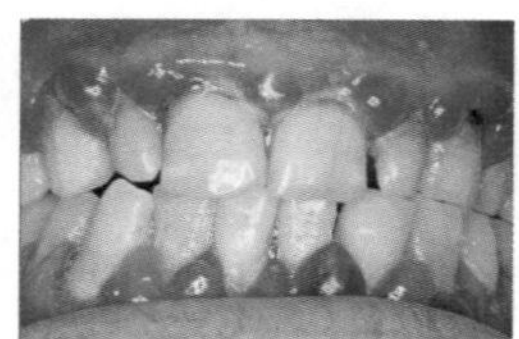
© BIOPHOTO ASSOCIATES/ SCIENCE PHOTO LIBRARY

Deficiency of vitamin C means that hydroxylase won't work properly and hydroxyproline is not formed. This means the hydrogen bonds can't form and the **connective tissue weakens**. This can lead to a condition called **scurvy** which causes bleeding gums, bleeding under the skin, bleeding around the joints and poor wound healing.

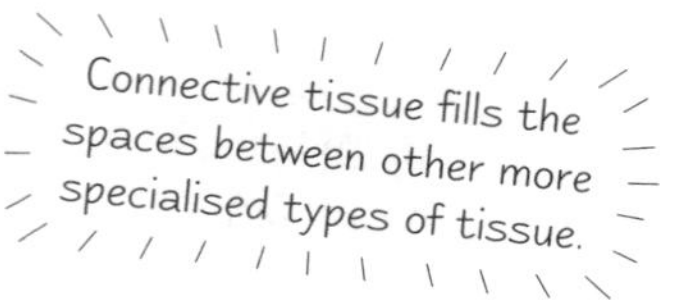

Vitamin D is needed for effective Absorption of Calcium

Vitamin D (calciferol) is a fat-soluble vitamin found in fish liver oil, eggs and butter. Also, when skin is exposed to ultra-violet light from the sun, it can make vitamin D from compounds related to cholesterol.

You need vitamin D to make a protein in the lining of the small intestine that binds to calcium ions in food. So vitamin D deficiency can lead to poor bone development — see opposite.

Balanced Diet

Calcium is needed for healthy development of Bones and Teeth

Calcium is found in **dairy** products, **bread** and **flour**. It's absorbed from the small intestine through a **protein carrier**. This carrier can only be made if vitamin D is present in the diet (see last page). In fact, people can suffer from the effects of calcium deficiency because of lack of vitamin D, even though there might be plenty of calcium in the diet. Calcium deficiency in the diet leads to **rickets** in children and fracture-prone bones in adults (**osteomalacia**).

Calcium is also needed for **nerve** and **muscle** function — it plays an important role at synapses (see page 26). This means that a deficiency in calcium (or vitamin D) can cause muscular spasms too.

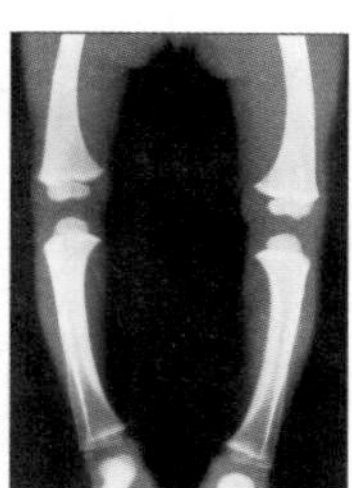

Bow legs are a common symptom of rickets.

Some chemicals (like **oxalate**) make calcium insoluble, which reduces its absorption. Spinach, for example, contains oxalate which reacts with calcium in this way. So while vitamin D is an **enhancer** of calcium absorption, oxalate is an **inhibitor**.

Iron is needed to make Haemoglobin

Iron is found in both meat and plants. It's needed for **haemoglobin** in red blood cells, so a deficiency of iron causes a form of **anaemia**, due to insufficient haemoglobin in the blood. Sufferers of anaemia are visibly paler and feel tired due to the fact that less oxygen is being carried to respiring cells.

Iron **absorption** is affected by various factors:

> Iron from meat is found in an organic molecule called **haem**, which normally exists bound to a protein, e.g. haemoglobin. This haem-iron is more **effectively absorbed** than the non-haem iron of plants, so you tend to get less iron from a **vegetarian diet**. But, absorption of haem-iron is still affected by various **enhancers** or **inhibitors** (similar to what happens with calcium). Absorption also depends on the amount of iron **already stored** — if this amount suddenly goes down, e.g. during menstruation, the body absorbs a lot more.

Iodine is needed for healthy growth

You get iodine in seafood. Iodine forms part of the hormone **thyroxine**.

Iodine deficiency in children causes poor growth and poor mental development — this is called **cretinism**. In adults, it causes **goitre**. One of the symptoms of goitre is a swelling in the neck region.

Practice Questions

Q1 List the four main groups of organic food compounds.

Q2 What is the role of vitamin C in the manufacture of collagen?

Q3 Explain why deficiency of vitamin D may lead to problems with bone development.

Q4 Name one essential mineral that is affected by absorption enhancers and inhibitors.

Q5 What are the possible effects of iodine deficiency in children and in adults?

Exam Questions

Q1 a) Explain why people with vitamin C deficiency have poor wound healing. [4 marks]

b) Suggest two good sources of vitamin C. [2 marks]

Q2 Explain why a strict vegetarian diet could lead to anaemia. [6 marks]

So carrots make you see in the dark...

That's not totally true, but you can make vitamin A from a chemical in carrots, and that helps your eyes. There are loads of vitamins and minerals here to learn. You need to know about them, or you might accidentally get nasty problems from not eating the right things — like the student who got scurvy from eating nothing but porridge. Yuck.

Effect of Lifestyle on Health

There's one basic rule in life — anything you enjoy is likely to be bad for your health. With the exception of exercise, but who really prefers jogging to pizza? Most things are OK in moderation, but if you overdo the fun, you pay for it. Boo

You Have to Watch Your **Diet**

As far as **diet** goes, the main problems are:

- eating **too much**
- having too much **fat** or **salt** in your diet.

Here are some of the nasty things that can happen (even the words are pretty nasty, but the effects are worse):

Atherosclerosis. This is the 'furring up' of the **coronary artery** by fats — mostly **cholesterol**. The fatty deposit inside the artery wall is called an **atheroma**, and it's caused by a diet that's too high in animal fats, which the body turns into cholesterol. This restricts the blood flow to the heart and can result in a **myocardial infarction** (heart attack, to you and me).

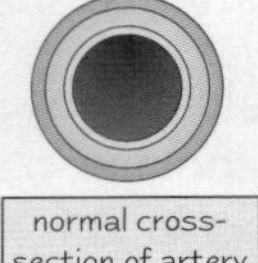

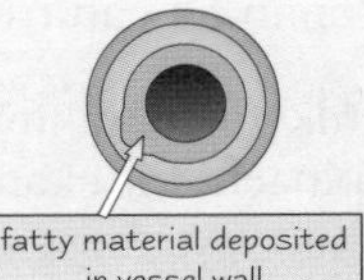

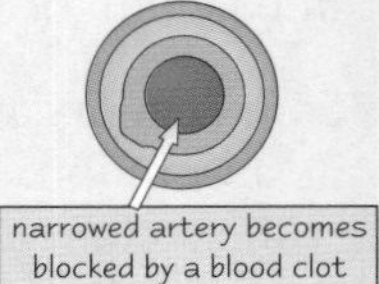

Atherosclerosis makes the formation of **clots** more likely, and if these block a blood vessel in the brain, a **stroke** results. When the blood supply to a part of the brain is cut off, that part gets damaged. The medical term for a stroke is a **cerebrovascular accident**.

Being overweight or eating too much fat or salt can result in **high blood pressure**. Of course, the medical term isn't 'high blood pressure' but **hypertension**. Hypertension makes your heart work harder and can cause heart attacks and strokes. It also increases the risk of developing an **aneurysm** (a bulge in an artery) which can be **fatal** if it bursts.

The best way to **reduce weight** and avoid all the nasty problems above is to **exercise more** and **eat less**. (Well, pretty much.) The type of exercise you do is important — fairly **vigorous exercise** is needed to reduce body fat. Weight should be lost **gradually** by reducing **energy-dense** foods (like carbohydrates and lipids). It's not a good idea to reduce **nutrient-dense** foods (such as those containing protein and vitamins) as it could lead to **deficiencies**.

There can be **Nasty Consequences** if you Don't Take Care of your **Lungs**

Here are some nasty diseases caused by mistreatment of your **lungs**.

1) **Lung cancer**. The main cause of lung cancer is **smoking** and you're also more likely to get it if you live or work in an atmosphere where other people smoke. **Tumours** grow in the epithelial tissue of the lungs and damage them, and tumour cells travel around the body in the blood and set up **secondary tumours** (**metastases**) elsewhere.

2) **Chronic bronchitis** is another disease that results from **smoking**, and it can also be caused by **air pollution**. Irritation causes the bronchioles in the lungs to become **inflamed**, which can block some of them. This makes breathing more difficult. The person coughs — which is an unsuccessful attempt to remove the blockage — and the **alveoli** (air sacs) in the lungs can **tear** as a result.

3) **Emphysema** is caused by smoking and air pollution too. It often follows if you've had chronic bronchitis for a long time. **Macrophages** digest the particles inhaled and produce **elastase**. The elastase digests the **elastin** in the alveoli wall, so they can't recoil any more. They burst, reducing the surface for gas exchange. The lungs **enlarge** but get **less efficient**. The person will be breathless and may even need to be given **oxygen** to help them breathe.

4) **Pneumoconiosis** is fibrosis and scarring of the lungs due to long-term breathing of dusts like coal, silica or asbestos. It's seen mainly in people who work in dusty environments, like coalmines or quarries.

5) **Tuberculosis** (**TB**) is an **infectious** disease caused by a bacterium, unlike the other four. Symptoms include a bad cough, fever and unexplained weight loss.

TREATMENT OF LIFESTYLE-RELATED HEALTH PROBLEMS

Disease	Treatment
atherosclerosis	surgical removal of section of artery, angioplasty (unblocking of arteries using a needle)
hypertension	low sodium diet and drug treatment (e.g. diuretics, beta blockers, calcium channel blockers), low fat diet and exercise programme
coronary heart disease	surgery, low fat diet, exercise programme, drug treatment (e.g. aspirin, statins, nitrates)
chronic bronchitis	drug treatment (e.g. bronchodilators, corticosteroids), oxygen therapy, lung transplant
tuberculosis (TB)	drug treatment (long-term course of several antibiotics)
pneumoconiosis	no treatment, except to avoid further inhalation
lung cancer	radiotherapy, chemotherapy, surgery

Effect of Lifestyle on Health

Holidays *may be* ***Good for You****, but excessive* ***Sunbathing Isn't***

Everyone likes sunlight (except vampires), but actually it's quite nasty. It contains damaging **ultra-violet radiation**. This radiation can cause **skin cancers** of various sorts, because it damages the **DNA** in the cells it hits. This can cause them to divide uncontrollably and become **malignant**. Ultra-violet (UV) radiation comes in various types:

1) **UVA** is what causes you to **tan** — it stimulates production of **melanin** (a protective brown pigment) in the skin. It penetrates deep into your skin and prolonged exposure can cause **skin cancer**.

2) **UVB** is what causes **sunburn**. It also causes **skin cancer**.

3) **UVC** doesn't usually get through the atmosphere — it's filtered out by the **ozone layer**. If it does get through, where the ozone layer is thin or absent (e.g. due to **CFCs**), it's very damaging to the skin.

Most skin cancers are fairly easy to treat and not too dangerous, particularly if they're caught early, but there's one very serious type called **malignant melanoma** which kills about **1600** people each year in the U.K. One reason it's so serious is that it can **spread** very quickly to other parts of the body if it's not treated at an early stage.

Protection against UV rays involves:

1) not exposing yourself to the sun for **long periods** (wear loose clothing and a sun hat instead of exposing your skin).
2) avoiding the **midday sun** (from around 11 am to 3 pm).
3) not using **sunbeds**.
4) using **sunscreen** of at least factor 15.

She looks happy now,
but a few years down the line...
...she'll realise how stupid that hat looked.

Practice Questions

Q1 What is an atheroma?

Q2 What is a cerebrovascular accident?

Q3 Name three lung diseases that can be caused by smoking.

Q4 Give one type of treatment for each of the diseases you named in your answer to Q3.

Q5 Which components of sunlight lead to skin cancer?

Exam Questions

Q1 Explain the health problems that can be caused by a diet too high in saturated fats. [6 marks]

Q2 Explain why sunbathing can be dangerous, and state ways in which these problems can be reduced. [10 marks]

TB or not TB, that is the question...

...hopefully not TB is the answer. These two pages aren't that bad really, but there are quite a few medical terms you have to learn, especially in that first section. Doctors like to sound clever, so they use words like 'myocardial infarction' instead of the much simpler 'heart attack', and you'll have to too.

Screening Programmes

Screening programmes are used to check whether people in high risk groups are developing diseases before they start to feel ill. This allows treatment to begin really early if a problem's found, which increases its chances of success.

Screening can Detect some Diseases Before any Symptoms are Seen

Some serious diseases have symptoms that can go **unnoticed** for a while, as they don't cause sickness, pain or discomfort. To detect these diseases early enough to allow for **successful treatment**, a variety of **screening programmes** have been introduced. A screening programme looks at as many people as possible, but especially those in some sort of **high risk category** (e.g. because of their age or because they have a family history of a disease), to detect those that have an early stage of a disease.

X-rays can be used to Detect Breast Cancer

A **breast x-ray** or **mammogram** is currently used to detect **breast cancer** in its earliest stages. Early detection can improve the chances of **successful treatment** and recovery. Breast x-ray screening aims to show changes which are **too small** to be felt by the patient and which can't be found in any other way. A **tumour** will show up on the x-ray as a **shadow**. Screening is normally carried out on women **over 50**, and is recommended once every **three years**. Radiation is **dangerous** to some extent (it can damage cells and lead to cancer), which is one reason why screening isn't done more often. It's done on women over 50 because they're **more at risk**, and in younger women the breast tissue tends to be more **dense** and x-rays can't pick up tumours very easily. It's estimated that the screening programme currently saves around **300 lives** per year in the UK.

Endoscopy Looks Inside the Body

1) **Endoscopy** involves passing small **optic fibres** into the body which are attached to a **camera** so a doctor can actually look inside your body for physical changes.
2) A miniature instrument can also be used to **snip off** a piece of tissue and retrieve it if the doctors want to test it.
3) Endoscopy isn't generally painful but it can be **uncomfortable**. People are unlikely to volunteer for it for no reason, so it's **not** used for general screening but only for investigating patients who have **some symptoms**.
4) It's commonly used to detect **cancer of the colon** (large intestine) and is also used to investigate **bowel irritations** and **lung cancer**.

Ultrasound is a Very Safe Screening Technique

Both **x-rays** and **endoscopy** carry a small **risk** for the patient. But the use of **ultrasound** to build up images of the body has no known side-effects or dangers. For this reason it's routinely used to screen **developing fetuses** during pregnancy.

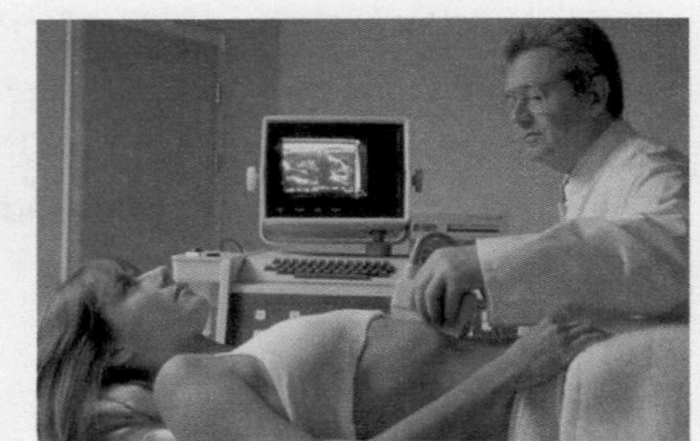

A doctor uses ultrasound to build up a sonogram on the monitor

Ultrasound is emitted from a **transducer** which is placed on the mother's abdomen and moved to "look at" different parts of the fetus. **Repetitive ultrasound beams** scan the fetus in thin slices and are **reflected back** onto the same transducer. The information obtained from different reflections are used to build up a picture (a **sonogram**) on a monitor screen. Sonograms are used to determine **age**, **size** and **growth** in the fetus, and to screen for any **problems** with its development.

Because of the **safety** of the procedure, ultrasound is being developed to screen for other conditions such as **blood clots**, **ovarian cancer**, and **breast cancer**. For breast cancer, it may be **more reliable** than x-rays for use in women **under 50**, as ultrasound can pick up irregularities in **denser** breast tissue than x-rays can.

Screening Programmes

Pedigree Analysis and **Genetic Screening** Tell us About **Inheritance**

Potential carriers of genetic diseases often want to know their **risk** of having children with those diseases. There are two ways they can find out about this risk:

- **Pedigree analysis** means using a **family tree** to see how genetic disorders in the family can be inherited. Understanding the **laws** of genetics helps work out the **probability** of offspring inheriting a particular gene.
- **Genetic screening** involves fairly complex techniques which can accurately tell whether there are **mutated** alleles present in unborn babies. Tests can be carried out on embryos **in vitro** (in glass, like a test-tube) or on fetuses **in utero** (in the uterus). Two examples of **in utero** testing are **amniocentesis** (taking samples of the amniotic fluid) and **chorionic villus sampling** (taking samples from the placenta).

Genetic Screening is a **Sensitive Issue**

Genetic screening may be suitable for **embryos** before implantation in **IVF treatment**, and for parents with a **known risk** of passing on a serious genetic condition to their children. The extraction of cells for genetic analysis carries some **risk** for both the mother and the fetus in a normal pregnancy, though, so it's **not** done routinely.

Other types of **genetic screening** involve testing children or adults for '**dormant**' **genetic diseases**, or testing potential parents to assess the risk that they will have a child with a serious genetic defect. This testing is **safe**, but has **ethical** considerations:

- It must always be **voluntary** — it's unethical to force people to undergo genetic screening.
- People are likely to need some **counselling** after being told that they're at risk.
- The information must be **confidential**.
- If information on health problems is known, **insurance companies** may demand that people tell them before quoting for insurance.

People who think they might be at risk of developing a genetic disease or of having a child with one can be referred to a **genetic counsellor**. A genetic counsellor has to have a sound knowledge of human genetics, and must be able to explain all the genetic ratios and probabilities in a way that their patients can understand. They also need to be able to deal sensitively with worried or distressed patients, and be able to explain all the **legal** and **medical options** — e.g. if a scan has revealed that a pregnant woman's baby has a serious genetic condition, the genetic counsellor would have to explain all the options to the parents, while ensuring that they are allowed to make their own choice.

Practice Questions

Q1 What is a mammogram?

Q2 Name two diseases that are screened for using endoscopy.

Q3 What name is given to a picture produced by ultrasound?

Q4 Give two reasons why ultrasound might be better than x-rays for screening women under 50 for breast cancer.

Exam Questions

Q1 Explain the reasons why screening for breast cancer is restricted to women over 50 and is only carried out once every three years. [5 marks]

Q2 Why is ultrasound preferred to x-rays when doctors want to see a picture of a developing fetus? [3 marks]

Q3 Huntington's disease is a dominant inherited disease involving the degeneration of nervous system cells, including brain cells, usually beginning at age 40+. The disease is fatal. Genetic testing can detect the disease, but there is no known treatment. Suggest **two** advantages and **two** problems that could result if a genetic screening programme for Huntington's disease was introduced. [4 marks]

The only screen I'm interested in is one with Coronation Street on it...

Just kidding. You can't fail to appreciate modern medical techniques that can save lives by detecting the very earliest stages of a serious illness. They could even save your life one day, so it shouldn't be too much of a hardship to learn a couple of pages on them. There's not that much to learn — the picture of the surprisingly trim pregnant woman takes up a lot of room.

Answers

Section 1 — Energy Supply

Page 3 — ATP and Energy Supply

1 Maximum of 2 marks available.
ATP is a molecule made from adenosine diphosphate (ADP) and phosphate, using energy from reactions like those of respiration **[1 mark]**. The energy stored in the chemical bond between the ADP and the phosphate can be released when it is needed by a cell, by breaking the ATP back down into ADP and phosphate **[1 mark]**.

2 Maximum of 3 marks available.
Because ATP is small and water-soluble, it can be easily transported around a cell or between cells to the places where there is a demand for energy **[1 mark]**. There it can be rapidly converted back into ADP to release the energy stored in the bonds **[1 mark]**. Because an enzyme is required for this reaction, there is little risk of the ATP breaking down into ADP and releasing its energy in the wrong place, wasting the energy **[1 mark]**.

Page 5 — Glycolysis and the Link Reaction

1 Maximum of 5 marks available.
The 6-carbon glucose molecule is phosphorylated using phosphate from 2 molecules of ATP **[1 mark]**, then hydrolysed / split using water **[1 mark]** to give 2 molecules of the 3-carbon molecule triose phosphate **[1 mark]**. This is then oxidised by removing hydrogen ions **[1 mark]** to give 2 molecules of 3-carbon pyruvate **[1 mark]**.

2 Maximum of 4 marks available.
The 3-carbon pyruvate is combined with coenzyme A **[1 mark]** to form a 2-carbon molecule, acetyl coenzyme A **[1 mark]**. The extra carbon is released as carbon dioxide **[1 mark]**. The coenzyme NAD is converted into reduced NAD in this reaction by accepting hydrogen ions **[1 mark]**.

Page 7 — Krebs Cycle and the Electron Transport Chain

1 Maximum of 14 marks available.
1 mark can be awarded for any of the following points, even if the final answer is incorrect:
2 ATP are produced in glycolysis **[1 mark]**. 1 ATP is produced per turn of the Krebs cycle **[1 mark]**, which happens twice per molecule of glucose **[1 mark]** giving 2 ATP from the Krebs cycle per molecule of glucose **[1 mark]**.
In the electron transport chain, 2.5 ATP are produced for every molecule of reduced NAD coenzyme made in the earlier stages of respiration **[1 mark]**, and 1.5 ATP for every molecule of reduced FAD produced **[1 mark]**.
2 reduced NAD are produced in glycolysis **[1 mark]**,
1 reduced NAD is produced in the link reaction **[1 mark]**
and 3 in the Krebs cycle **[1 mark]**, but for every molecule of glucose, 2 molecules of pyruvate are made by glycolysis **[1 mark]**, so the link reaction and Krebs cycle happen twice per molecule of glucose **[1 mark]**.
So in total, 8 molecules of reduced NAD are produced by the link reaction and the Krebs cycle **[1 mark]**. Adding the 2 reduced NAD produced in glycolysis gives 10 molecules of reduced NAD **[1 mark]**.
$10 \times 2.5 = 25$ ATP **[1 mark]**.
1 molecule of reduced FAD is also produced per turn of the Krebs cycle **[1 mark]**, giving 2 reduced FAD per glucose molecule **[1 mark]**.
$2 \times 1.5 = 3$ ATP **[1 mark]**.
So in total, the electron transport chain produces $25 + 3 = 28$ ATP **[1 mark]**. Adding the ATP produced in glycolysis and in the Krebs cycle gives $28 + 2 + 2 = 32$ molecules of ATP in total **[1 mark]**.

Page 9 — Photosynthesis

1 a) In the thylakoid membranes of the chloroplasts **[1 mark]**.
b) Maximum of 2 marks available.
ATP **[1 mark]** and NADPH + H^+ / reduced NADP / NADPH **[1 mark]**.

Page 11 — Photosynthesis

1 Maximum of 5 marks available.
a) Ribulose bisphosphate (RuBP) **[1 mark]**.
b) NADPH / reduced NADP / NADPH + H^+ **[1 mark]**.
c) Ribulose bisphosphate (RuBP) **[1 mark]**.
d) The enzyme ribulose bisphosphate carboxylase **[1 mark]**.
e) Ribulose bisphosphate (RuBP) **[1 mark]**.

2 Maximum of 3 marks available.
a) Between points a and b, light was available and both stages of photosynthesis (light-dependent and light-independent) were happening. ATP and NADPH / reduced NADP / NADPH + H^+ were being supplied for the Calvin cycle **[1 mark]**.
b) At point b the light faded and the light-dependent stage of photosynthesis stopped, but the light-independent stage / Calvin cycle continued until point c **[1 mark]**.
c) Photosynthesis stopped at c as supplies of ATP and NADPH were exhausted and no more could be produced **[1 mark]**.

Section 2 — Control, Coordination and Homeostasis

Page 13 — Homeostasis and Temperature Control

1 a) Maximum of 2 marks available.
A change in a factor brings about a response that counteracts the change / makes it opposite **[1 mark]** so that the factor returns to a norm **[1 mark]**.
b) Maximum of 2 marks available from any of the following.
Body temperature **[1 mark]**, blood glucose concentration **[1 mark]**, water potential **[1 mark]**. Or other sensible answer.

Page 15 — Removal of Metabolic Waste

1 a) Maximum of 5 marks available.
Microvilli provide a large surface area of membrane **[1 mark]** so there is a large surface available for substances to pass across and more carrier proteins **[1 mark]**. There are lots of mitochondria **[1 mark]** which provide energy / ATP for reabsorption **[1 mark]** by active transport **[1 mark]**.
b) Maximum of 3 marks available for any of the following:
More glucose passes into the blood, because it is actively reabsorbed **[1 mark]**, but there is no carrier protein for urea **[1 mark]**. Some urea passes into the blood by diffusion **[1 mark]** because it's a small molecule **[1 mark]**.

Page 17 — Removal of Metabolic Waste

1 Maximum of 10 marks available.
Strenuous exercise causes more sweating **[1 mark]** so more water is lost from the body **[1 mark]**. This increases the blood solute concentration / decreases blood water potential / makes blood water potential more negative **[1 mark]**. It also stimulates osmoreceptors **[1 mark]** in the hypothalamus **[1 mark]**, which stimulates the pituitary gland **[1 mark]** to release **more** ADH **[1 mark]**. ADH increases the permeability of the collecting ducts **[1 mark]** so more water is reabsorbed into the blood by osmosis **[1 mark]**. This means that less water is lost in the urine which prevents further dehydration **[1 mark]**.

2 Maximum of 5 marks available.
More sodium chloride is removed from the ascending limb **[1 mark]** of a longer loop by active transport **[1 mark]** so solute concentration rises / solute potential falls / water potential falls / becomes more negative in the medulla **[1 mark]**; so **more** water is reabsorbed from the collecting duct **[1 mark]** by osmosis **[1 mark]**.

Page 19 — Stimulus and Response

1 a) Maximum of 3 marks available for any of the following points:
Receptors can communicate with effectors by releasing a chemical / hormone that binds to them **[1 mark]**. The chemical / hormone moves from the receptor to the effector by diffusion if the two cells are close together **[1 mark]**, or if they are far apart the chemical / hormone is released into the blood and transported via mass flow **[1 mark]**. The other main way that receptors can send a message to

Answers

effectors is to trigger an electrical impulse in a nerve, which is then passed through the nervous system until it reaches the effector and triggers a response **[1 mark]**.

b) Maximum of 2 marks available.
Plants don't have nervous systems, and have to release a chemical in order to respond to changes in their environment **[1 mark]**. This is much slower as it has to rely on diffusion **[1 mark]**.

2 Maximum of 4 marks available for any of the following points:
Chemoreceptor cells have membrane-bound receptor molecules / proteins **[1 mark]**. The stimulus molecule must have a complementary shape **[1 mark]** to bind to the receptor molecule **[1 mark]** and create a generator potential **[1 mark]** to transmit a nerve impulse **[1 mark]**.

Page 21 — Regulation of Blood Glucose

1 Maximum of 10 marks available from the following.
Glucose is absorbed into the blood (from the gut) **[1 mark]** which makes the blood glucose concentration increase **[1 mark]**.
This stimulates the beta cells **[1 mark]** of the islets of Langerhans **[1 mark]** to secrete insulin **[1 mark]**. Insulin is released into the bloodstream **[1 mark]** and binds to receptors **[1 mark]** on liver cells **[1 mark]**. This increases the permeability of the liver cells to glucose **[1 mark]**, converting glucose into glycogen / stimulating glycogenesis **[1 mark]**. This reduces blood glucose concentration **[1 mark]**.

2 Maximum of 5 marks available.
Exercise uses up glucose (by respiration) **[1 mark]** which lowers blood glucose concentration **[1 mark]** and stimulates the alpha cells **[1 mark]** of the islets of Langerhans **[1 mark]** to secrete glucagon **[1 mark]**.

Page 23 — The Mammalian Eye

1 Maximum of 8 marks available.
Cones are responsible for the high acuity of the human eye **[1 mark]**. They're densely packed at the fovea, where most of the light that enters the eye tends to focus **[1 mark]**. Each synapses with just one bipolar neurone **[1 mark]**, so it can send very detailed information to the brain **[1 mark]**. Rods are found in the more peripheral parts of the retina **[1 mark]**. They're more sensitive than cones, because it takes less light to activate the pigment inside them **[1 mark]**. Lots of rods converge onto the same bipolar neurone too **[1 mark]**, so small responses from each one can combine to send a message to the brain through that bipolar neurone **[1 mark]**.

Page 25 — The Nerve Impulse

1 a) Stimulus **[1 mark]**.

b) Maximum of 3 marks available.
A stimulus causes sodium channels in the neurone cell membrane to open **[1 mark]**. Sodium **diffuses** into the cell **[1 mark]**, so the membrane becomes depolarised / more positive charge moves in (than moves out) **[1 mark]**.

c) Maximum of 2 marks available, from any of the following.
The membrane was in the refractory period **[1 mark]** and so the sodium channels were inactive / recovering / couldn't be opened **[1 mark]**. Alternatively, the stimulus could have been lower than threshold level **[1 mark]**.

2 Maximum of 5 marks available, for any 5 of the following:
Transmission of action potentials will be slower **[1 mark]**. Myelin insulates the axon / has high electrical resistance **[1 mark]** and there are gaps / nodes of Ranvier between sheaths **[1 mark]** where depolarisation happens /sodium channels are concentrated **[1 mark]**. So in an intact myelinated axon, saltatory transmission occurs / action potentials jump from node to node **[1 mark]**. This can't happen if the myelin sheath is damaged / more membrane is exposed **[1 mark]**.

Page 27 — Synapses and Synaptic Transmission

1 Maximum of 8 marks available, for any 8 of the following:
Arrival of the action potential causes calcium channels to open (in the presynaptic membrane) **[1 mark]** which makes calcium diffuse into the bouton / synaptic knob / cell **[1 mark]**. This stimulates the vesicles in the bouton to fuse with the presynaptic membrane **[1 mark]** and release the neurotransmitter **[1 mark]** by exocytosis **[1 mark]**. The neurotransmitter diffuses across the gap / cleft **[1 mark]** and binds to receptors on the postsynaptic membrane **[1 mark]**. This stimulates opening of sodium channels (on the postsynaptic membrane) **[1 mark]** so sodium diffuses into the cell **[1 mark]**. The membrane then becomes depolarised **[1 mark]**.

2 Maximum of 4 marks available.
Vesicles (containing neurotransmitter) are only found in the presynaptic neurone **[1 mark]**, so exocytosis / release / secretion (of neurotransmitter) can only happen from here **[1 mark]**. The receptors (for the neurotransmitter) are only found on the postsynaptic membrane **[1 mark]** so only this membrane can be stimulated by the neurotransmitter / neurotransmitter can only bind here **[1 mark]**.

Page 29 — The Brain

1 Maximum of 8 marks available from any of the following.
Action potentials / nerve impulses arrive at primary sensory areas **[1 mark]**. There are different sensory areas for different parts of the body **[1 mark]** — the left hemisphere receives impulses from the right side of the body, and vice-versa **[1 mark]**. The more sensory cells (in a part of the body), the bigger the sensory area of the cerebrum **[1 mark]**. Association areas integrate information **[1 mark]** — they pass action potentials / nerve impulses to motor neurones **[1 mark]**, so they control the responses **[1 mark]**. Examples of association areas are Broca's area (involved in speech) **[1 mark]** and Wernicke's area (language) **[1 mark]**.

2 Maximum of 2 marks available.
The auditory motor area (Broca's area) is damaged **[1 mark]** because this is the area that would normally coordinate muscle movement to produce speech **[1 mark]**.

Page 31 — The Autonomic Nervous System

1 a) Maximum of 2 marks available.
It's involuntary **[1 mark]** and stereotypic / always the same kind of response **[1 mark]**.

b) Maximum of 2 marks available.
Different muscles have different receptors **[1 mark]** which trigger different chemical effects inside the muscles **[1 mark]**.

Page 33 — Muscle Structure and Function

1 Maximum of 10 marks available, for any 10 of the following:
The sarcoplasmic reticulum membranes become much more permeable **[1 mark]** and calcium ions diffuse out **[1 mark]**. They reach the actin filaments and bind to a protein called troponin **[1 mark]**, which causes another protein called tropomyosin to change position **[1 mark]** and unblock the binding sites on the actin filaments **[1 mark]**. The myosin heads attach to the binding sites **[1 mark]** to form actomyosin cross bridges between the two filaments **[1 mark]**. The myosin head then changes angle **[1 mark]**, pulling the actin over the myosin towards the centre of the sarcomere **[1 mark]**. The cross bridges then detach and reattach further along the actin filament **[1 mark]**. ATP provides the energy for this **[1 mark]**.

2 Maximum of 5 marks available.
More sarcoplasmic reticulum enables action potentials to be carried to more myofilaments **[1 mark]** so more calcium can be released into the cytoplasm **[1 mark]**. This means that there is more stimulation of the actin-myosin interaction **[1 mark]**, so more of the muscle / more myofibrils are stimulated to contract **[1 mark]**. This gives more efficient / faster / stronger muscle contraction **[1 mark]**.

Answers

Section 3 — Inheritance

Page 35 — Meiosis

1 *Maximum of 2 marks available.*
*Ovaries **[1 mark]** and testes **[1 mark]**.*

2 *Maximum of 4 marks available.*
*a) A gene is a section of DNA that controls one characteristic / controls the synthesis of one or more polypeptide(s) **[1 mark]**. An allele is one of the alternative forms of a gene **[1 mark]**.*
*b) Haploid cells have one complete set of chromosomes **[1 mark]**, diploid cells have two complete sets of chromosomes — in pairs of homologous chromosomes **[1 mark]**.*

3 *Maximum of 3 marks available.*
*Meiosis halves the chromosome number **[1 mark]**, so it compensates for the doubling of chromosome number at fertilisation **[1 mark]**. Meiosis also increases variety by producing new combinations of alleles **[1 mark]**.*

Page 37 — Inheritance

1 *Maximum of 3 marks available — 1 mark for every two correct answers from the following.*
I^AI^A, I^AI^O, I^BI^B, I^BI^O, I^AI^B, I^OI^O

2 *Maximum of 3 marks available.*
*Carry out a test cross **[1 mark]**, by crossing with a white-flowered plant **[1 mark]**. If some of the offspring are white-flowered then the plant was heterozygous / If all purple-flowered then the plant was homozygous. **[1 mark]***

Page 39 — Inheritance

1 *Maximum of 4 marks available.*

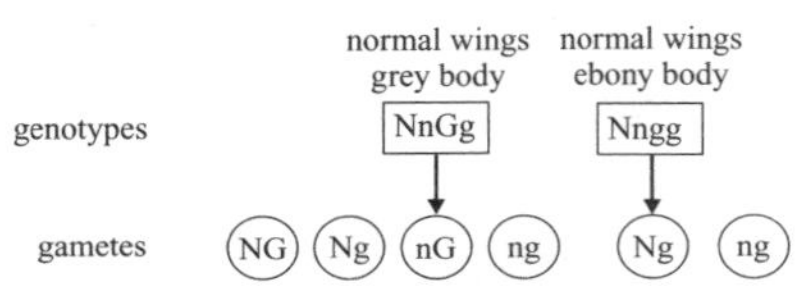

	NG	Ng	nG	ng
Ng	NNGg normal, grey	NNgg normal, ebony	NnGg normal, grey	Nngg normal, ebony
ng	NnGg normal, grey	Nngg normal, ebony	nnGg vestigial, grey	nngg vestigial, ebony

Ratio of 3 normal wings and grey bodies : 3 normal wings and ebony bodies : 1 vestigial wings and grey body : 1 vestigial wings and ebony body.
4 marks for all genotypes and phenotypes correct.
Deduct one mark for each mistake.

2 *Maximum of 5 marks available for any of the points below.*
*The recessive allele for haemophilia is carried on the X chromosome, so males will only have one copy of the allele and females will have two copies **[1 mark]**. This means that a female who inherits one copy of the haemophilia allele will also probably have a copy of the 'normal' allele for factor 8, which is dominant **[1 mark]**. She will be healthy, but a carrier **[1 mark]**. Males who inherit one copy of the haemophilia allele will have the disease **[1 mark]**, so they can get it if they have a healthy father and a carrier mother **[1 mark]**. Females only stand a chance of having haemophilia if they're the child of a haemophiliac male and a carrier female **[1 mark]**, so the probability is much lower **[1 mark]**.*

Page 41 — Variation

1 *Maximum of 2 marks available.*
*2^3 or $2 \times 2 \times 2$ **[1 mark]**, so 8 possibilities **[1 mark]**.*
In a question like this, always show your working.

2 *a) Characteristics that show continuous variation are more likely to be affected by the environment **[1 mark]**.*
*b) Characteristics that show continuous variation usually involve more genes **[1 mark]**.*

Page 43 — Natural Selection and Evolution

1 *Maximum of 6 marks available from the following points:*
*Natural selection has occurred **[1 mark]**. Before the drought individual birds showed variation in beak depth and length **[1 mark]**. When the drought happened birds with longer and deeper beaks had an advantage because they were able to obtain food that the others could not **[1 mark]**. The birds with the longest and deepest beaks had the greatest chance of survival **[1 mark]**. The birds who survived produced offspring with larger beaks **[1 mark]**. The alleles for larger beaks would therefore be more prominent in future generations of* Geospiza fortis *on the island **[1 mark]**. This is an example of directional selection **[1 mark]**.*

Page 45 — Speciation

1 a) *Maximum of 3 marks available.*
*Darwin noticed that there were 14 different species of finch on the Galápagos islands **[1 mark]**. Each of these finches occupied a different ecological niche **[1 mark]**. The finches had different beaks which were adapted to eating different foods **[1 mark]**.*
b) *Maximum of 4 marks available.*
*Darwin thought that all the finches were originally one species living on one island and competing for resources **[1 mark]**. Some finches flew to other islands and established separate populations **[1 mark]**. Adaptation to different habitats / food sources gradually changed the beak shapes of the finches **[1 mark]**. This eventually led to the formation of new species **[1 mark]**.*

Page 47 — Classification

1 *Maximum of 2 marks available.*
*Traditional classification deals with features that are easy to observe **[1 mark]** whilst phylogeny deals with the genetic relationships between organisms **[1 mark]**.*

2 *Maximum of 5 marks available.*
*a) phylum **[1 mark]**, b) class **[1 mark]**, c) family **[1 mark]**, d) Aptenodytes **[1 mark]**, e) patagonicus **[1 mark]**.*

Section 4 — Environment

Page 49 — Ecosystems and Energy Transfer

1 *Maximum of 4 marks available.*
*A habitat is the place where an organism or group of organisms live **[1 mark]**. An ecosystem is not just the place, it also includes other abiotic factors **[1 mark]** such as temperature, oxygen level, soil pH, exposure to wind, etc. **[1 mark for a relevant example]** and biotic factors / living things / communities that live there **[1 mark]**.*

2 *Maximum of 4 marks available.*
*Pyramids of number are a different shape when a single large organism can feed many smaller organisms **[1 mark]**. Pyramids of biomass are a different shape if you have a population of short-lived and rapidly reproducing organisms **[1 mark]**. Pyramids of energy can never be out of shape because, due to energy wastage between trophic levels **[1 mark]**, there always has to be more energy in the population being fed on than the population feeding on it **[1 mark]**.*

Page 51 — Nutrient Cycles

1 *Maximum of 10 marks available.*
*Carbon dioxide is removed from the atmosphere by photosynthesis **[1 mark]**. Carbon dioxide is returned to the atmosphere by respiration **[1 mark]**. Carbon dioxide not released by respiration is released when the organism dies **[1 mark]** by the respiration of decomposers **[1 mark]**. If organisms do not decay, their carbon is not released **[1 mark]**. This carbon can eventually form fossil fuels **[1 mark]**. Humans burn large quantities of fossil fuels **[1 mark]** which releases a lot of carbon dioxide **[1 mark]**. Deforestation can also raise carbon dioxide levels **[1 mark]** by killing trees that would absorb it **[1 mark]**.*

Answers

2 Maximum of 6 marks available.
Nitrogen-fixing bacteria convert nitrogen into amino acids **[1 mark]**. *This is important because the amino acids may either be absorbed directly by plants* **[1 mark]**, *or may be decomposed later to nitrates that plants can absorb* **[1 mark]**. *Microbial decomposers break down the bodies of dead plants and animals and release the nitrogen* **[1 mark]** *as ammonium compounds* **[1 mark]**. *The ammonium compounds are converted into nitrites and then to nitrates by nitrifying bacteria* **[1 mark]**. *Denitrifying bacteria convert nitrates into nitrogen* **[1 mark]**.

Page 53 — Studying Ecosystems

1 Maximum of 2 marks available.
A transect would be used to discover a trend across an ecosystem **[1 mark]**. *Any suitable example (e.g. distribution of organisms up a rocky shore, distribution of plants with increasing shade)* **[1 mark]**.

2 Maximum of 8 marks available.
Map the field **[1 mark]**. *Divide it into numbered squares* **[1 mark]**. *Use a random sampling technique to select the squares to sample* **[1 mark]**. *Place quadrats in the selected squares* **[1 mark]**. *Quadrats should be divided into 100 smaller squares* **[1 mark]**. *Count the number of squares of clover* **[1 mark]**. *Count only squares which are at least half occupied by clover* **[1 mark]**. *Average the count for all the quadrats used OR add the total number of squares covered in all the quadrats and convert to a percentage of the total number of squares sampled* **[1 mark]**.

Page 55 — Studying Ecosystems

1 Maximum of 3 marks available.
The fertiliser makes no significant difference **[1 mark]**.
The probability value is more than 0.1 **[1 mark]**. *This means that the difference between the crops would occur by chance more than one time in 10 / 10% of the time, which is too often to be able to reject the null hypothesis* **[1 mark]**.

Page 57 — Dynamics of Ecosystems

1 Maximum of 4 marks available.
A density dependent factor has more effect as population density increases **[1 mark]**. *A relevant example (e.g. food, oxygen, minerals etc.)* **[1 mark]**. *A density independent factor's intensity is unaffected by the density of a population* **[1 mark]**. *A relevant example (e.g. fire, flood, drought etc.)* **[1 mark]**.

Page 59 — Succession

1 Maximum of 8 marks available.
Succession is a process where plant communities gradually develop on bare land **[1 mark]**. *Change goes through stages called seral stages* **[1 mark]**. *The process stops when a climax community is reached* **[1 mark]**. *Changes are brought about by the interactions of species* **[1 mark]**. *Climax may be climatic* **[1 mark]** *resulting from the climate* **[1 mark]** *or a plagioclimax* **[1 mark]** *resulting from human activity* **[1 mark]**.

2 a) Maximum of 3 marks available from the points below:
Successful features — rapid growth **[1 mark]**, *rapid reproduction* **[1 mark]**, *asexual reproduction* **[1 mark]**, *efficient seed dispersal* **[1 mark]**, *tolerant of harsh environmental conditions e.g. high salt levels and strong winds* **[1 mark]**.

b) Maximum of 2 marks available.
Reasons for disappearance — shaded by larger plants **[1 mark]**, *eaten by herbivores* **[1 mark]**, *unable to compete for water or minerals with newly arrived species* **[1 mark]**.

Page 61 — Agriculture and Ecosystems

1 Max of 6 marks available from the following points:
(at least 2 must be for benefits).
Benefits — Efficient food production **[1 mark]**, *increased food production* **[1 mark]**, *cheaper food* **[1 mark]**, *takes up less space* **[1 mark]**.
Problems — Pollution from waste **[1 mark]**, *increased use of pesticides* **[1 mark]**, *increased use of inorganic fertilisers* **[1 mark]**, *problem with increased animal waste* **[1 mark]**, *destruction of hedgerows* **[1 mark]**, *unemployment of farm workers* **[1 mark]**, *reduces biodiversity of farm land* **[1 mark]**.

2 Maximum of 5 marks available from the points below:
Advantages — Improves soil structure **[1 mark]**, *less likely to pollute ponds and streams* **[1 mark]**, *uses animal waste* **[1 mark]**.
Disadvantages — more difficult to store **[1 mark]**, *more difficult to apply to land* **[1 mark]**, *can't be measured out so easily* **[1 mark]**.
Manure can have an unpleasant smell **[1 mark]**.

3 Maximum of 8 marks available.
Eutrophication results from pollution of fresh water with fertilisers **[1 mark]**. *Fertilisers leach through the soil from farmland and get into the water* **[1 mark]**. *Nitrates* **[1 mark]** *in the fertilisers increase the growth of algae* **[1 mark]**. *This blocks out the light and plants below then die* **[1 mark]**. *This causes bacterial growth as bacteria feed on the dead plants* **[1 mark]**. *Bacteria use up oxygen* **[1 mark]**. *Lack of oxygen kills organisms in the pond or stream* **[1 mark]**.

Section 5 — Applied Ecology

Page 63 — Diversity

1 Maximum of 2 marks available.
Population size $= n_1 \times n_2 / n_m = 80 \times 100 / 10 = 800$
[2 marks for correct answer or 1 mark for correct working].

Page 65 — Pollution of Aquatic Ecosystems

1 Maximum of 2 marks available.
The water is probably polluted with organic matter **[1 mark]** *and has a low oxygen concentration* **[1 mark]**.

2 Maximum of 2 marks available.
The presence of the E. coli *bacterium* **[1 mark]** *would indicate that the water is polluted with human sewage. This is because* E. coli *is usually found in the human large intestine* **[1 mark]**.

3 Maximum of 3 marks available from any 3 of the following.
Freshwater shrimps **[1 mark]**, *stonefly larvae* **[1 mark]**, *mayfly larvae* **[1 mark]** *and caddis fly larvae* **[1 mark]**.

Page 67 — Adaptations to the Environment

1 a) Maximum of 3 marks available.
Smaller shrews have a larger surface area:volume ratio **[1 mark]**, *so they lose heat more rapidly* **[1 mark]**. *They need to burn / respire more food to replace this heat energy* **[1 mark]**.

b) Maximum of 2 marks available.
Shrews / mammals are endothermic **[1 mark]**, *so they lose more heat to their surroundings (than ectothermic insects)* **[1 mark]**.

Page 69 — Agricultural Ecosystems and Crop Production

1 Maximum of 3 marks available.
In the tropical rainforest there are more plants (meaning a higher productivity) because there are more mineral nutrients available **[1 mark]** *and there is more water available* **[1 mark]**. *The plants in the tropical rainforest also have a higher leaf area index, which allows them to photosynthesise more efficiently — desert plants have small leaves or spines to conserve water* **[1 mark]**.

2 a) Maximum of 2 marks available.
Not all the available light is absorbed **[1 mark]**, *because there aren't enough leaves. Some light that could have been absorbed will pass straight through the leaves or miss them altogether* **[1 mark]**.

b) Maximum of 2 marks available.
Leaves in the lower layers don't get enough light because they're blocked by the leaves above **[1 mark]**. *They use more food in respiration than they produce in photosynthesis, so energy is wasted* **[1 mark]**.

Answers

Page 71 — Harvesting Ecosystems

1 *Maximum of 4 marks available.*
Advantages of fishing — any 2 of the following 3 points:
*An extra food resource, over and above farming **[1 mark]**. Allows a natural ecosystem to continue alongside food production **[1 mark]**. Encourages governments etc. to manage conservation of fish stocks **[1 mark]**.*
Disadvantages of fishing — any 2 of the following 3 points:
*Overfishing can destroy the fish population **[1 mark]**. Fishing can disrupt the marine food web **[1 mark]**. The natural ecosystem can't be controlled like a farm can **[1 mark]**.*

2 *Maximum of 2 marks available.*
*Antibiotics could kill beneficial bacteria **[1 mark]**, and overuse of the antibiotics might lead to antibiotic resistance in the bacteria **[1 mark]**.*

Page 73 — Conservation of Species

1 *Maximum of 2 marks available.*
*The human population has increased **[1 mark]**, and industrialisation, agriculture, deforestation and other practices damaging to the environment have increased to support it **[1 mark]**.*

2 *Maximum of 4 marks available the following points:*
*Visitors may damage habitats — perhaps just by repeatedly walking through them **[1 mark]**. Visitors may disturb animals / pick plants **[1 mark]**. Visitors may pollute / leave litter **[1 mark]**. Humans aren't part of the natural ecosystem **[1 mark]**. It may be necessary to limit the number of visitors or their access to certain areas in order to reduce this damage **[1 mark]**.*

Page 75 — Controlling Pests

1 *Maximum of 3 marks available from any of the following:*
*Non-toxic to humans **[1 mark]**. Specific to weed species being targeted **[1 mark]**. Biodegradable / not persistent **[1 mark]**. Easy to wash off / remove from crops before consumption **[1 mark]**.*

2 *Maximum of 5 marks available from any of the following:*
*Biological control is more difficult to apply **[1 mark]**. Biological control can be unpredictable — it's hard to predict all of the knock-on effects **[1 mark]**. Biological control is slower to work than chemical pesticides **[1 mark]**. Biological control does not control sudden outbreaks as well as pesticides do **[1 mark]**. Biological control does not completely eliminate the pest **[1 mark]**. Switching from chemical pesticides to biological control will be costly / the farmer may have to purchase new supplies and equipment **[1 mark]**.*

Section 6 — Microbes and Disease

Page 77 — Bacteria

1 *Maximum of 4 marks available.*
*Genetic recombination can occur by conjugation **[1 mark]**, by exchanging plasmids **[1 mark]**, by transfer in bacteriophage viruses **[1 mark]**, or by taking up DNA from the surroundings **[1 mark]**.*

2 *Maximum of 2 marks available.*
*The layer of polysaccharide might prevent desiccation / drying out **[1 mark]** and attack from antibodies / phagocytes / white blood cells **[1 mark]**.*

Page 79 — Culturing Microorganisms

1 a) *Maximum of 5 marks available.*
*Hold a wire loop in a flame until red hot and then cool **[1 mark]**. Ensure that there is minimal lifting of the lids from the monoculture and sterile plate **[1 mark]**. Dip the loop into the monoculture **[1 mark]**, then streak the surface of the medium **[1 mark]**. One mark for any other precaution, e.g. work close to bunsen flame, wear protective clothing, swab bench with disinfectant.*

b) *Maximum of 2 marks available.*
*Heat in an autoclave **[1 mark]** to 121°C for 15 minutes **[1 mark]**.*
Also accept:
The agar plates would be irradiated using gamma radiation.

Page 81 — Bacterial Growth

1 *Maximum of 8 marks available.*
*Take a known / fixed volume of culture at the start time **[1 mark]**. Dilute it by adding water of a known volume / dilute by a known amount **[1 mark]**. Repeat the dilution a fixed number of times **[1 mark]**. Spread a known volume of the final dilution on an agar plate **[1 mark]**. Incubate it **[1 mark]**, then count the number of colonies formed **[1 mark]**. Multiply this number by each dilution factor **[1 mark]**. Repeat for the sample taken at the end time **[1 mark]**.*

2 a) *Maximum of 3 marks available.*
*The bacteria respired aerobically **[1 mark]**, which released more energy / made more ATP **[1 mark]** for cell growth / division **[1 mark]**.*

b) *Maximum of 2 marks available.*
*There would be no population growth / cells would die **[1 mark]** because oxygen is poisonous to them **[1 mark]**.*

Page 83 — Industrial Growth of Microorganisms

1 a) *Maximum of 2 marks available for any of the following.*
*Batch culture occurs in fixed volume, continuous culture doesn't **[1 mark]**. Medium is added to the continuous culture vessel at a constant rate, but none is added to a batch culture after its start **[1 mark]**. Culture reaches the end of the stationary phase in batch culture, but it's kept at the exponential phase in continuous culture **[1 mark]**. Batch cultures have to be started over again regularly, unlike continuous cultures which last much longer **[1 mark]**.*

b) *Maximum of 2 marks available.*
*Advantages of batch culture: easier to control conditions / less costly to start again in event of contamination **[1 mark]**. Advantages of continuous culture: method is more productive / culture lasts for longer / a smaller vessel can be used, linked with stated advantage (e.g. easier to sterilise) **[1 mark]**.*

c) *Maximum of 3 marks available. Possible answers include:*
Penicillium *would get less oxygen **[1 mark]**, so the culture would respire less **[1 mark]**, so less energy / less growth would occur **[1 mark]**, and less fungus / mycelium produces less penicillin **[1 mark]**.*

Page 85 — Bacterial Disease

1 *Maximum of 4 marks available. Possible answers include:*
Salmonella *bacteria bind to the (epithelium of) the gut wall **[1 mark]**, where they are taken up by phagocytosis **[1 mark]**. This causes inflammation **[1 mark]**.* Salmonella *bacteria contain endotoxins **[1 mark]**, and their release prevents absorption of water in the large intestine, leading to diarrhoea **[1 mark]**.*

2 *Maximum of 3 marks available.*
Staphylococcus *bacteria remain in the gut cavity **[1 mark]** where they release exotoxins **[1 mark]**. Their release prevents absorption of water in the large intestine, resulting in diarrhoea **[1 mark]**.*

Page 87 — Viral Disease

1 *Maximum of 2 marks available.*
*The latency period is the time between the initial invasion of a cell and the replication of the virus **[1 mark]** when the virus is dormant **[1 mark]**.*

2 *Maximum of 3 marks available.*
*Without the enzyme reverse transcriptase, DNA cannot be made from RNA **[1 mark]**; so DNA cannot be inserted into the host chromosome **[1 mark]**; so the virus cannot reproduce **[1 mark]**.*

Page 89 — Protection Against Disease

1 *Maximum of 3 marks available.*
Lysosomes fuse with the phagocytic vesicles containing the engulfed

Answers

bacteria or material, giving phagolysosomes ***[1 mark]****. The lysosomes contain hydrolytic enzymes* ***[1 mark]****, hydrogen peroxide and free radicals to kill bacteria and break down organic material* ***[1 mark]****.*

2 *Maximum of 2 marks available.*
As capillaries become more leaky, phagocytes / neutrophils squeeze through pores in the capillary wall ***[1 mark]*** *to reach the site of infection* ***[1 mark]****.*

3 *Maximum of 3 marks available.*
An antigen is a foreign particle ***[1 mark]*** *that sets up an immune response* ***[1 mark]****. It is recognised by the body as non-self* ***[1 mark]****.*

Page 91 — Cell and Antibody Mediated Immunity

1 a) *Maximum of 5 marks available.*
An antigen binds to membrane-bound antibodies on the surface of B-lymphocytes ***[1 mark]****. This stimulates mitosis / cell division of the lymphocytes to produce clones* ***[1 mark]****. These lymphocytes then produce and secrete / release antibodies* ***[1 mark]*** *that can bind to / are complementary to the antigen* ***[1 mark]*** *making the antigen harmless* ***[1 mark]****.*
b) *Maximum of 1 mark available.*
In the bone marrow and thymus when B- and T-lymphocytes are maturing, any that have receptors that fit self antigens are destroyed ***[1 mark]****.*

Page 93 — Antibiotics and Vaccines

1 a) *One from:*
More people were getting vaccinated / The vaccine may take a while to become effective ***[1 mark]****. (Or other sensible answer.)*
b) *One from:*
The polio virus is still present / there are still a small number of cases / the disease could be re-introduced from outside / the virus might be 'dormant' in the population ***[1 mark]****.*

2 *Maximum of 4 marks available. Possible answers include:*
A bacterium in the population has a random mutation ***[1 mark]*** *that causes it to produce the enzyme penicillinase* ***[1 mark]****. This blocks the action of penicillin* ***[1 mark]****. The mutant bacteria are more likely to survive and reproduce when exposed to penicillin* ***[1 mark]****. The gene for resistance is passed to their offspring* ***[1 mark]*** *and also transferred by conjugation via plasmids* ***[1 mark]****.*

Section 7 — Behaviour and Populations

Page 95 — Simple Behaviour Patterns

1 *Maximum of 3 marks available from the following (accept other relevant examples, e.g. imprinting in ducklings).*
Human babies are born with the innate (genetic) ability to speak ***[1 mark]****. However, they have to learn language* ***[1 mark]****. Which language they learn depends on their environment* ***[1 mark]****— Japanese babies learn Japanese, but a Japanese baby brought up by German-speakers would learn German* ***[1 mark]****.*

2 *Maximum of 3 marks available.*
Operant conditioning could be used in dog training to reward ***[1 mark]*** *or punish specific behaviours* ***[1 mark]****. E.g. the dog could be given a biscuit each time it offered a paw to its owner, and it would learn to perform this behaviour to receive a biscuit* ***[1 mark]****.*

Page 97 — Courtship and Territory

1 *Maximum of 3 marks available.*
Stereotyped courtship behaviour ensures that any potential mate is of the right species ***[1 mark]****. It ensures that mating only occurs if the female is in a receptive condition / mating will produce offspring* ***[1 mark]****. It also allows the female to select a mate with features that will give her young a good chance of survival if inherited (this could be explained in a variety of ways e.g. strongest, best display etc. — the key idea is that the female has a basis for choice)* ***[1 mark]****.*

2 a) *Maximum of 2 marks available. Possible answers include:*
Pheromones travel long distances ***[1 mark]****, so the female can attract males over a wide area with little effort* ***[1 mark]*** *and select the best of them for her mate* ***[1 mark]****.*
b) *Maximum of 1 mark available, for any one of the following:*
Females can attract several males by producing pheromones, and can then choose between them to select the strongest alleles for her offspring ***[1 mark]****. / If lots of females were attracted to a single male, some of them might not reproduce at all and fewer offspring would be produced overall* ***[1 mark]****. / It's better for females not to travel too far and compete for males, but to conserve their energy for producing and raising young* ***[1 mark]****. / Females only produce pheromones when they're ready for breeding — it would be pointless for males to attract females that were not ready to mate* ***[1 mark]****.*

3 *Maximum of 5 marks available.*
Having a territory allows the breeding pair exclusive use of the resources in an area ***[1 mark]****. This means that their young are more likely to survive* ***[1 mark]****. Certain features (usually in the male) lead to successfully acquiring / defending a territory* ***[1 mark]****. Better survival of the young means that genes for successful territorial behaviour continue in the offspring* ***[1 mark]****. Over several generations, territorial behaviour will become widespread in the population because of its selective advantage* ***[1 mark]****.*

Page 99 — Hormonal Control of Reproduction

1 a) *Maximum of 3 marks available.*
A = progesterone ***[1 mark]****.*
B = prolactin ***[1 mark]****.*
C = HCG ***[1 mark]****.*
b) *Maximum of 2 marks available.*
Prolactin is produced in the anterior pituitary gland ***[1 mark]****, and stimulates milk production* ***[1 mark]****.*

Page 101 — Conception, Contraception and Infertility

1 *Maximum of 4 marks available.*
Injection of extracted or synthetic gonadotrophin hormones ***[1 mark]****. Gonadotrophins stimulate ovum / secondary oocyte development and release* ***[1 mark]****. Treatment using the drug clomiphene* ***[1 mark]****. Clomiphene stimulates natural gonadotrophin production* ***[1 mark]****.*

2 *Maximum of 7 marks available.*
A drug is taken to stimulate superovulation / production of several ova ***[1 mark]****. Sperm are collected from the male* ***[1 mark]****. Eggs are collected from the female's ovary using a needle and ultrasound* ***[1 mark]****. The eggs and sperm are incubated overnight* ***[1 mark]****. The fertilised eggs are selected and developed under laboratory conditions for a few days* ***[1 mark]****. The fertilised eggs are then transferred into the mother's uterus* ***[1 mark]****. The mother receives progesterone treatment throughout procedure, so that the uterus is suitable for implantation* ***[1 mark]****.*

Page 103 — Pregnancy

1 a) *Maximum of 12 marks available, for any 12 of the following:*
Oxygen passes from mother to fetus ***[1 mark]*** *by diffusion down a concentration gradient* ***[1 mark]****. This movement is helped by the higher affinity of fetal haemoglobin for oxygen* ***[1 mark]****. Carbon dioxide passes from fetus to mother* ***[1 mark]*** *by diffusion down a concentration gradient* ***[1 mark]****. Nutrients move from mother to fetus* ***[1 mark]****. Glucose moves across via facilitated diffusion* ***[1 mark]****, amino acids by active transport* ***[1 mark]****, and vitamins and minerals by a combination of diffusion and active transport* ***[1 mark]****. Water moves from mother to fetus* ***[1 mark]*** *by osmosis* ***[1 mark]****. Urea moves from fetus to mother* ***[1 mark]*** *by diffusion* ***[1 mark]****. Some antibodies can pass from mother to fetus across the placenta* ***[1 mark]****, as can nicotine, alcohol and other drugs* ***[1 mark]****.*
b) *Maximum of 2 marks available.*
The placenta acts as an endocrine organ / produces hormones / secretes oestrogen, progesterone and HCG during pregnancy ***[1 mark]****. The placenta acts as a barrier to bacteria and some viruses* ***[1 mark]****.*

Answers

2 *Maximum of 6 marks available, for any 6 of the following: Pregnant women put on weight due to: The weight of the baby* ***[1 mark]****. The weight of the placenta* ***[1 mark]****. An increased volume of blood / body fluids* ***[1 mark]*** *to provide extra energy for the demands of pregnancy and breast feeding* ***[1 mark]****. Breast development / growth* ***[1 mark]*** *ready for breast feeding* ***[1 mark]****. Increased fat storage* ***[1 mark]*** *to provide energy for the demands of pregnancy and breast feeding* ***[1 mark]****.*

Page 105 — Growth and Development

1 a) *0 – 1 years* ***[1 mark]****.*
b) *6 cm/year* ***[1 mark]****.*
c) *17 years* ***[1 mark]****.*

2 *Maximum of 8 marks available for any 8 of the following: Older people tend to be less mobile because their tissues begin to degenerate* ***[1 mark]*** *and the structural proteins become harder and less elastic* ***[1 mark]****. This makes them stiffer and less supple* ***[1 mark]****. They also experience more health problems as their organs begin to function less effectively* ***[1 mark]****, and their immune system becomes less efficient* ***[1 mark]****. Accuracy of DNA replication also declines with age* ***[1 mark]****, which can lead to the production of abnormal cells which don't work properly* ***[1 mark]****, or die* ***[1 mark]****, or become cancerous* ***[1 mark]****.*

Page 107 — Population Size and Structure

1 *Maximum of 5 marks available. The population would decrease in size* ***[1 mark]****. There are a large number of older people* ***[1 mark]*** *who are too old / unlikely to have children* ***[1 mark]****. There are not many young people* ***[1 mark]*** *to have children in the future* ***[1 mark]****. As an alternative to saying that there are a lot of old people / not many young people, it would be acceptable to say the pyramid is wide at the top / narrow at the base.*

2 *Maximum of 4 marks available. In the low density population, deer tend to live longer / death rate increases sharply when the deer reach old age / deer tend to die from natural causes as they near maximum life expectancy* ***[1 mark]****. In the high density population, deer have a more constant death rate / die at a younger age more often* ***[1 mark]****. Deer tend to die at a younger age in the high density population due to greater competition for food* ***[1 mark]*** *and increased likelihood of disease or predation* ***[1 mark]****.*

Page 109 — Infectious Disease and Immunity

1 a) *One from: More people were getting vaccinated / The vaccine may take a while to become effective* ***[1 mark]****. (Or other sensible answer.)*
b) *One from: The polio virus is still present / there are still a small number of cases / the disease could be re-introduced from outside / the virus might be 'dormant' in the population* ***[1 mark]****.*

Page 111 — Balanced Diet

1 a) *Maximum of 4 marks available. Vitamin C acts as a cofactor / coenzyme for the enzyme needed to make collagen* ***[1 mark]****. The enzyme converts proline into hydroxyproline* ***[1 mark]*** *so that hydrogen bonds can form (between hydroxyproline units / residues)* ***[1 mark]****. Collagen needs to be replaced for wounds to heal* ***[1 mark]****.*
b) *Maximum of 2 marks available. Citrus fruits* ***[1 mark]****, green vegetables* ***[1 mark]****, potatoes* ***[1 mark]****. Also, allow separate marks for named examples of citrus fruits.*

2 *Maximum of 6 marks available. Less iron in a vegetarian diet* ***[1 mark]****. Iron in a vegetarian diet is non-haem* ***[1 mark]****, which is absorbed less effectively than iron from meat / haem iron* ***[1 mark]****. This means less iron is absorbed into the blood* ***[1 mark]****, so less haemoglobin is made* ***[1 mark]****. Less haemoglobin leads to anaemia* ***[1 mark]****.*

Page 113 — Effect of Lifestyle on Health

1 *Maximum of 6 marks available. The fats are converted into cholesterol* ***[1 mark]****. Cholesterol is deposited in the artery walls and causes atheroma / atherosclerosis* ***[1 mark]****. Blood flow through the arteries is reduced* ***[1 mark]****. This can put a strain on the heart* ***[1 mark]*** *and / or can cut off the blood supply to an area of the heart causing a heart attack / myocardial infarction* ***[1 mark]****. Reduced blood flow to the brain can cause localised brain damage resulting in a stroke / cerebrovascular accident* ***[1 mark]****.*

2 *Maximum of 10 marks available, for any 10 of the following: Sunbathing exposes the skin to UV light* ***[1 mark]****. This penetrates the skin* ***[1 mark]*** *and can damage the DNA in a cell* ***[1 mark]****. Sometimes the genes that control cell division are damaged* ***[1 mark]****, and then the cell undergoes uncontrolled mitosis and tumours develop* ***[1 mark]****. The cells are said to be cancerous / malignant* ***[1 mark]*** *leading to skin cancer* ***[1 mark]****. The danger can be reduced by: Avoiding exposure to the sun for long periods / covering up the skin with clothing and / or hats* ***[1 mark]****. Avoiding the midday sun* ***[1 mark]****. Not using sunbeds* ***[1 mark]****. Using sunscreen of at least factor 15* ***[1 mark]****.*

Page 115 — Screening Programmes

1 *Maximum of 5 marks available, for any 5 of the following: X-rays carry a risk of damaging cells / causing cancer / causing mutations* ***[1 mark]****, so patients shouldn't be exposed to them too often* ***[1 mark]****. Women under 50 are much less likely to develop breast cancer, so the small risk involved in having a mammogram isn't as worthwhile* ***[1 mark]****. Women under 50 have denser breast tissue* ***[1 mark]*** *and so x-rays don't work as well for them* ***[1 mark]****. Screening programmes can be expensive, so only the women most at risk are tested* ***[1 mark]****.*

2 *Maximum of 3 marks available. Ultrasound is harmless* ***[1 mark]****. X-rays can cause mutation of cells* ***[1 mark]****, which would be particularly dangerous in a developing fetus as it has relatively few cells, and they're dividing to form important structures* ***[1 mark]****.*

3 *Maximum of 4 marks available. Advantages — Patients and their families would be prepared for the onset of the disease and could plan how best to cope* ***[1 mark]****. They could avoid passing the disease on to any children* ***[1 mark]****. Disadvantages — The person would know they had a fatal disease, perhaps many years before they died, which would cause a lot of stress and anxiety* ***[1 mark]****. The person could be refused life insurance* ***[1 mark]****. The person might be refused employment* ***[1 mark]****. The person's family life and relationships could be badly affected* ***[1 mark]****.*
(Any 2 of the disadvantages listed, for a maximum of 2 marks.)

Index

Index

Index

Index